Gauge Theory in Higher Dimensions

Daniel G. Fadel
Henrique N. Sá Earp

Gauge Theory in Higher Dimensions

INSTITUTO DE MATEMÁTICA PURA E APLICADA

Gauge Theory in Higher Dimensions

ISBN   978-85-244-0480-1
MSC (2010) Primary: 53C38, Secondary: 97I10, 53C07, 57R57, 58E15, 81T13

Coordenação Geral   Nancy Garcia

Produção   Books in Bytes   **Capa**   Sergio Vaz

Realização da Editora do IMPA
   IMPA   Telefones: (21) 2529-5005
   Estrada Dona Castorina, 110   2529-5276
   Jardim Botânico   www.impa.br
   22460-320 Rio de Janeiro RJ   editora@impa.br

```
SA127g    Sá Earp, Henrique
              Gauge Theory in Higher Dimensions / Daniel G. Fadel e
          Henrique N. Sá Earp. - 1.ed. -- Rio de Janeiro: IMPA, junho de
          2019.
              ISBN 978-85-244-0480-1
              1. Gauge theory. 2. Higher dimension. I. Título. II. Série.
              CDD: 516  CDU: 514
```

Contents

This book is an introduction to gauge theory in dimensions greater than 4. It is organised as a reader's guide to Gang Tian's landmark article *Gauge theory and calibrated geometry, I*, published in the *Annals of Mathematics*, Tian (2000). That paper carries out to substantial lengths the programme outlined by Simon Donaldson and Richard Thomas in the seminal paper Donaldson and Thomas (1998), by extending the fundamental notion of *instanton* in Yang–Mills theory to higher dimensional special geometries. Moreover, his *bubbling theorem* relates instanton compactness to the theory of calibrated submanifolds, the celebrated work by Harvey and Lawson (1982). This work expresses our admiration for all of these key authors.

The text is aimed at graduate students and advanced undergraduates, as well as specialists in other areas of Mathematics and Physics, with an interest in modern Differential Geometry. We adopt a fast-paced but self-contained exposition of the background on connections and curvature on bundles, special geometric structures, analysis on manifolds and geometric measure theory, alongside the proof of Tian's bubbling theorem. We also highlight some links with other important works in contemporary Geometry and Topology.

This version was commissioned as supporting material to an advanced course by the authors at the 32nd Brazilian Colloquium of Mathematics, hosted by the Institute for Pure and Applied Mathematics (IMPA) from July 29th to August 2nd, 2019. The contents are almost entirely adapted from the first-named author's Masters dissertation Fadel (2016), at the University of Campinas (Unicamp).

We hope that the reader will share in our enthusiasm for this beautiful and fast-evolving subject.

A brief history of gauge theory.

The advent of Yang–Mills theory in the mid-1970s had a strong influence on the development of differential geometry and topology Donaldson (2005). The basic concept in the theory is the *Yang–Mills functional*, defined on bundle connections ∇ over a given Riemannian manifold as the square of the L^2-norm of the curvature:

$$\mathcal{YM}(\nabla) := \| F_\nabla \|_{L^2}^2.$$

Its critical points are characterised by a second-order condition on ∇ called the *Yang–Mills equation*: $\mathrm{d}_\nabla^* F_\nabla = 0$. In four-dimensions, an important type of manifest solutions consists of so-called *(anti-)selfdual* connections, satisfying

$$*F_\nabla = \pm F_\nabla,$$

under the Hodge star operator. This is a first-order condition, which implies the Yang–Mills equation. The space of equivalence classes of such solutions, modulo symmetries, is called the *(A)SD instanton moduli space* (respectively). In particular, outstanding results on topology of 4–manifolds derive from the study of moduli spaces of anti-selfdual (ASD) instantons.

Building upon analytical works of Taubes (1982) and Uhlenbeck (1982a,b), Donaldson (1983) was able to show that certain intersection forms could not be realised by compact, simply-connected smooth 4–manifolds. One year earlier, Freedman (1982) had classified all compact, simply-connected *topological* 4–manifolds, so that Donaldson's result automatically gave several examples of previously unknown nonsmoothable 4–manifolds. Later on, Taubes (1987) proved a generalization of Donaldson's theorem for oriented asymptotically periodic 4–manifolds. This implied the existence of *uncountably* many *exotic* smooth structures on $\mathbb{R}^4$, i.e. the existence of an uncountable family of diffeomorphism classes of oriented 4–manifolds homeomorphic to $\mathbb{R}^4$ (see also the earlier work of Gompf (1985)). Ultimately, Donaldson extended his work significantly and produced deep new invariants distinguishing smooth 4–manifolds with the same intersection form Donaldson (1990), Donaldson and Kronheimer (1990), and Freed and Uhlenbeck (1984).

In the late-90s, the hugely influential work by Donaldson and Thomas (1998) outlined profound insights towards extending the theory to *higher dimensions* (i.e. greater than 4), in the presence of special geometric structures. Such a generalisation of the notion of instanton was first considered by physicists in Corrigan et al. (1983) and further discussed in Baulieu, Kanno, and Singer (1998) and Carrión (1998). While the classification of differentiable structures is much better understood in higher dimensions Scorpan (2005), one may optimistically expect gauge theory to lead to new invariants for special holonomy manifolds, such as Calabi–Yau, G_2 and Spin(7) manifolds.

In order to carry out the Donaldson–Thomas program rigorously, one would like to have higher-dimensional analogues of the compactness results by Uhlenbeck. In fact, the possible failure of compactness is a major issue: for any sequence $\{\nabla_i\}$ of Yang–Mills connections on a $G-$bundle $E \to M$, with uniformly bounded L^2-energy $\|F_{\nabla_i}\|^2_{L^2} \leqslant \Lambda$, there exists a closed 'blow-up set' $S \subseteq M$, of Hausdorff codimension at least 4, such that, up to gauge transformations, a subsequence of ∇_i converges to a Yang–Mills connection in $C^\infty_{\mathrm{loc}}-$topology away from S Nakajima (1988), Price (1983), and Uhlenbeck (n.d.).

Instanton bubbling and calibrated geometry.

Tian (2000) initiated this programme by proving foundational regularity results concerning blow-up loci of general sequences of Yang–Mills connections, notably showing that these are rectifiable, admit a natural geometric structure, and that at each point of this subset the sequence loses energy via bubbling and possibly also develops a singularity. Tian's analysis is similar to the work of Lin (1999) on the analogous compactness problem for harmonic maps, and his key tools are Price's monotonicity formula Price (1983) and a curvature estimate due to Uhlenbeck and Nakajima (1988).

The paper begins with a very general formulation of anti-selfduality, for connections on a $G-$bundle E over an oriented Riemannian manifold (M, g) endowed with a closed $(n - 4)-$form Ξ. A connection ∇ is called a $\Xi-anti$-*selfdual instanton* if

$$*(\Xi \wedge F_\nabla) = -F_\nabla.$$

This first-order condition still implies the Yang–Mills equation. Moreover, when the manifold M is closed, each $\Xi-$ASD instanton has an a priori L^2-energy bound, depending only on E, M and Ξ. For suitable choices of Ξ, the $\Xi-$ASD equation generalises the familiar ASD equations in $4-$dimensions, the Hermitian Yang–Mills equations on Kähler manifolds, and the higher-dimensional equations of G_2- and Spin(7)$-$instantons (in particular, complex ASD instantons). Tian's major breakthrough was the discovery of a specific relation between gauge theory and calibrated geometry: when Ξ is a calibration, the blow-up set of a sequence of $\Xi-$ASD instantons defines a $\Xi-calibrated$ *integer rectifiable current* Tian (2000, Theorem 4.3.2), and thus a (possibly very singular) $\Xi-$calibrated submanifold.

The present text offers a comprehensive treatment of Tian's Ξ-instanton bubbling theory, specifically Chapters $1 - 4$ of Tian (ibid.), which we organise in the following two theorems:

Theorem A (Uhlenbeck, Price, Nakajima, Tian). *Let (M^n, g) be a connected, closed, oriented Riemannian manifold, with $n \geq 4$. Let $E \to M$ be a G−bundle with compact structure group, and let $\{\nabla_i\}$ be a sequence of smooth Yang–Mills connections on E with uniformly bounded energy $\mathcal{YM}(\nabla_i) := \|F_{\nabla_i}\|^2_{L^2} \leq \Lambda$. Then, after passing to a subsequence, the following assertions hold:*

(i) *There exist a closed subset $S \subseteq M$ with finite Hausdorff measure $\mathcal{H}^{n-4}(S) < \infty$, a smooth Yang–Mills connection ∇ on $E|_{M \setminus S}$, and a sequence of gauge transformations $\{g_i\} \subseteq \mathcal{G}(E|_{M \setminus S})$, such that $g_i^* \nabla_i$ converges to ∇ in C^∞_{loc}−topology on $M \setminus S$.*

(ii) *There exist a constant $\varepsilon_0 > 0$, depending only on (M^n, g) and G, and a bounded upper semi-continuous function $\Theta : S \to [\varepsilon_0, \infty[$ such that, as Radon measures,*

$$\mu_i := |F_{\nabla_i}|^2 dV_g \rightharpoonup \mu = |F_\nabla|^2 dV_g + \Theta \mathcal{H}^{n-4} \lfloor S.$$

(iii) *S decomposes as $S = \Gamma \cup \operatorname{sing}(\nabla)$ with*

$$\Gamma := \operatorname{supp}(\Theta \mathcal{H}^{n-4} \lfloor S) \quad \text{and}$$

$$\operatorname{sing}(\nabla) := \left\{ x \in M : \limsup_{r \downarrow 0} r^{4-n} \int_{B_r(x)} |F_\nabla|^2 dV_g > 0 \right\};$$

Γ is countably $\mathcal{H}^{n-4}$−rectifiable, i.e. $T_x \Gamma$ is well-defined at $\mathcal{H}^{n-4}$−a.e. $x \in \Gamma$, and $\operatorname{sing}(\nabla)$ is $\mathcal{H}^{n-4}$−negligible.

(iv) *If $x \in \Gamma$ is a smooth point, i.e. $T_x \Gamma$ exists and $x \notin \operatorname{sing}(\nabla)$, then there is a non-flat Yang–Mills connection $I(x)$ on $T_x \Gamma^\perp$, with $\mathcal{YM}(I(x)) \leq \Theta(x)$, whose pull-back to $T_x M$ is gauge-equivalent to the limit of a blowing-up of $\{\nabla_i\}$ around x.*

Theorem B (Tian). *Let (M^n, g) be a connected, closed, oriented Riemannian manifold, with $n \geq 4$. Let Ξ be a smooth calibration $(n-4)$−form on M, i.e. Ξ is closed and has comass bounded by 1:*

$$\||\Xi|^*\|_{L^\infty(M)} \leq 1,$$

where

$$|\Xi|^*_x := \sup \{ \langle \Xi_x, v_1 \wedge \ldots \wedge v_{n-4} \rangle : v_i \in T_x M, |v_i| = 1 \}.$$

Let $E \to M$ be an $SU(r)$−*bundle over* M, *and let* $\{\nabla_i\}$ *be a sequence of smooth* Ξ−*anti-selfdual instantons. Then one has the* a priori *energy bound* $\mathcal{YM}(\nabla_i) = \langle c_2(E) \cup [\Xi], [M] \rangle$, *so that Theorem A holds, and furthermore we have the following:*

(i) *The limit connection* ∇ *is also a* Ξ−*anti-selfdual instanton away from* S.

(ii) *For any smooth point* $x \in \Gamma$, *the* $(n-4)$−*form* $\Xi_x := \Xi|_{T_x M}$ *restricts to one of the volume forms induced by g on $T_x \Gamma$, and $I(x)$ is a non-trivial ASD instanton on $(T_x \Gamma^\perp, g|_{T_x \Gamma^\perp})$ with respect to the orientation given by $*\Xi_x|_{T_x \Gamma^\perp}$. Equivalently, $B(x)$ is a Ξ_x−ASD instanton on $(T_x M, g|_{T_x M})$.*

(iii) *The* $(n-4)$−*current given by*

$$C(\Gamma, \Theta)(\varphi) := \frac{1}{8\pi^2} \int_\Gamma \langle \varphi, \Xi|_\Gamma \rangle \Theta \, \mathrm{d}\left(\mathcal{H}^{n-4} \lfloor \Gamma\right), \quad \forall \varphi \in \Omega^{n-4}(M),$$

is a Ξ−*calibrated integer rectifiable current satisfying conservation of instanton charge density, in the following sense: for every $\varphi \in \Omega^{n-4}(M)$,*

$$\lim_{i \to \infty} \int_M \mathrm{tr}(F_{\nabla_i} \wedge F_{\nabla_i}) \wedge \varphi = \int_M \mathrm{tr}(F_\nabla \wedge F_\nabla) \wedge \varphi + 8\pi^2 C(\Gamma, \Theta)(\varphi).$$

$$\tag{B1}$$

In particular, the L^2−*energy is conserved:*

$$\langle c_2(E) \cup [\Xi], [M] \rangle = \mathcal{YM}(\nabla) + \int_\Gamma \Theta \, \mathrm{d}\left(\mathcal{H}^{n-4} \lfloor \Gamma\right).$$

Remark. In the situation of Theorem B:

- The function Θ measures the energy density lost by the sequence around a point $x \in \Gamma$. If, instead of a single ASD instanton bubbling off transversely at $x \in \Gamma$, there is actually a whole bubbling tree of ASD instantons, then the inequality in (iv) of Theorem A is necessarily strict.

- Tao–Tian Tao and Tian (2004) further show that ∇ extends to a Ξ−anti-selfdual instanton on a G−bundle $\tilde{E}$ over $M \setminus \mathrm{sing}(\nabla)$ which is isomorphic to E over $M \setminus S$.

- In the simplest case, the singularities of ∇ are removable, Γ is a smooth Ξ−calibrated submanifold, and the bubbling trees of ASD instantons consist of single ASD instantons forming a smooth section I of an instanton bundle associated to the normal bundle of Γ and the restriction $E|_\Gamma$ of the ambient bundle. Conjecturally, in case (M, g, Ξ) is a G_2- or $\mathrm{Spin}(7)-$manifold, I should satisfy a certain non-linear Dirac equation associated to Ξ and the restriction $\nabla|_\Gamma$ called the *Fueter equation*, see Haydys (2017) and Walpuski (2017a,b).

Overview of the book.

In Chapter 1 we introduce the terminology of connections on $G-$bundles, their curvatures, and some related differential operators. This includes a classical Bochner–Weitzenböck formula Bourguignon and Lawson Jr (1981) for the generalised Hodge–de Rham Laplacian on $\mathfrak{g}_E-$valued 2–forms, and a review on Sobolev spaces of connections (Section 1.1). We also provide some material on holonomy groups and basic Chern–Weil theory (Sections 1.2 and 1.3). Next, we explain the weak and strong formulations of the Yang–Mills equation over Riemannian manifolds, discuss some of its symmetries and give a brief account of gauge fixing (Section 1.4). We illustrate the chapter with a basic survey of 4–dimensional gauge theory, reviewing classical (anti-)selfdual instantons, topological energy bounds via Chern–Weil theory, and their relation to Kähler geometry (Section 1.5).

In Chapter 2, we introduce the language of calibrated geometry, following Harvey and Lawson (1982), and the notion of instanton in higher dimensions, from Baulieu, Kanno, and Singer (1998), Corrigan et al. (1983), Donaldson and Thomas (1998), and Tian (2000). Since these notions naturally arise in the context of manifolds with special holonomy, we begin with a brief discussion of Riemannian holonomy groups and Berger's classification theorem, as well as some important geometries associated to special groups, such as Kähler for $\mathrm{U}(m)$, Calabi–Yau for $\mathrm{SU}(m)$, and the exceptional cases G_2 and $\mathrm{Spin}(7)$ (Section 2.1). Next, a quick review of minimal submanifolds motivates the introduction of calibrations and calibrated submanifolds, highlighting the classical examples on special holonomy manifolds (Section 2.2). Finally, we present two approaches for the generalization of the familiar 4–dimensional notion of instanton. The first approach is that of Tian (2000), based on the presence of a closed differential form $\Xi \in \Omega^{n-4}(M^n)$ (Section 2.3). The second approach, originally introduced by Carrión (1998), is formulated in terms of an $N(H)-$structure on M^n, where $N(H)$ is the normaliser of some closed Lie subgroup $H \subseteq \mathrm{SO}(n)$. Both points of view are shown to coincide in cases of interest, drawing various analogies between such instantons and calibrated submanifolds.

Chapter 3 establishes the analytical backbone of bubbling theory, leading up to the proof of claim (i) in Theorem A. We review Uhlenbeck's compactness theorems Uhlenbeck (1982a,b) (Section 3.1), and we study two core results in the analysis of Yang–Mills fields: a monotonicity formula, by Price (1983) (Section 3.2), and a local pointwise estimate with sufficiently small normalised L^2–norm over small balls, by Nakajima (1988) and Uhlenbeck (1982b) (Section 3.3). In particular, these results are used to show that sequences of Yang–Mills connections with uniformly bounded L^2–energy are C^∞_{loc}–convergent away from a blow-up set of Hausdorff codimension at least 4, along which the normalised L^2–energy concentrates (Section 3.4).

Finally, Chapter 4 addresses the structure of blow-up loci of sequences of Yang–Mills connections with uniformly bounded L^2–energy. The main results correspond to claims (ii)–(iv) in Theorem A and Theorem B. Fixing such a sequence $\{\nabla_i\}$, with limit connection ∇ away from the blow-up set S (subsequentially and modulo gauge), we will see that S decomposes into two closed pieces: one, denoted by Γ, which involves energy loss, and $\mathrm{sing}(\nabla)$, on which the formation of singularities occurs. The latter is readily shown to be a $\mathcal{H}^{n-4}$–negligible set (Section 4.1). Next, we show a first regularity result: Γ is (countably) $\mathcal{H}^{n-4}$–*rectifiable*, i.e. the approximate $(n-4)$–dimensional tangent space $T_x\Gamma$ of Γ exists, for $\mathcal{H}^{n-4}$–almost every $x \in \Gamma$ (Section 4.2). Then we move on to the behavior of ∇_i, for i sufficiently large, near a smooth point $x \in \Gamma$, at which $T_x\Gamma$ is well-defined. Applying blow-up analysis techniques that go back to Lin (1999), and following the more recent approach Walpuski (2017c), we can find a non-flat Yang–Mills connection $I(x)$ on $T_x\Gamma^\perp$ satisfying the energy inequality $\mathcal{YM}(I(x)) \leqslant \Theta(\mu, x)$ and whose pull-back to $T_x M$ is gauge-equivalent to the limit of a blowing up of the sequence $\{\nabla_i\}$ near x (Section 4.3).

Then we turn to the case in which $\{\nabla_i\}$ is a sequence of Ξ–anti-selfdual instantons (Section 4.4). We show that, for any $x \in \Gamma$, the approximate tangent space $T_x\Gamma$ is calibrated by Ξ, and an ASD instanton 'bubbles off' transversely; indeed $I(x)$ is a (non-flat) ASD instanton. Finally, we conclude the proof of Theorem B by showing that, for $G \subseteq \mathrm{SU}(r)$, the natural $(n-4)$–current $C(\Gamma, \Theta)$, defined by the triple $(\Gamma, \Xi, \frac{1}{8\pi^2}\Theta)$, is a Ξ–calibrated integer rectifiable current satisfying (B1).

Original contributions.

While our exposition follows Tian (2000) very closely, in the course of writing we took a few opportunities to streamline the argument, in the light of more recent literature, aimed at improving the reader's experience. Of course we bear full responsibility for those deviations, and the author of the source paper is in absolutely no way liable for any shortcomings deriving from our whim. Let us explain the main points at which our account diverges from the source.

In Sections 3.2-3.4 and Chapter 4 we adopt the general framework of n-manifolds of bounded geometry up to order 1, to which the monotonicity formula, ϵ-regularity, convergence outside the blow-up set, and the actual blow-up analysis can be carried over. Besides compact manifolds, this includes symmetric spaces and manifolds with noncompact ends, such as asymptotically cylindrical (ACyl) and asymptotically conical (AC) manifolds. Over the past decade, the latter geometries have attracted significant interest in higher-dimensional gauge theory, e.g. Sá Earp (2015) and Sá Earp and Walpuski (2015) and Jacob and Walpuski (2018), and Clarke and Oliveira (2019) and Lotay and Oliveira (2018), respectively. In particular, the proof of ϵ-regularity and the blow-up analysis follow a slightly more direct approach, formulated by Walpuski (2017c).

Finally, we stop short of reproducing in this exposition some claims from the original source. In particular, the claim that the current on M defined by

$$8\pi^2 c_2(\nabla) = \mathrm{tr}(F_\nabla \wedge F_\nabla) - \mathrm{tr}(F_\nabla) \wedge \mathrm{tr}(F_\nabla)$$

is closed Tian (2000, Corollary 2.3.2), which turns out to be equivalent to the assertion that $C(\Gamma, \Theta)$ is closed Tian (ibid., Theorem 4.2.3 (3)); in general this may not be true, see Petrache and Rivière (2017, §1.12.4). Moreover, we do not reproduce Tian's original proof that Θ takes values in $8\pi^2\mathbb{Z}$ in the Ξ−instantons case. Instead, we refer the reader to the more general energy identity result of A. Naber and Valtorta (2016) which easily implies the claim (see Theorem 4.35 and the discussion preceding Theorem 4.39).

1 Geometry and gauge theory

We begin by reviewing the basic terminology on $G-$bundles, connections and curvature, as well as some important associated differential operators over Riemannian manifolds, such as the Hodge-de Rham and the *rough* Laplacians, in Section 1.1. We proceed to two special topics that will be needed in Chapter 3: a corresponding Bochner–Weitzenböck formula and Sobolev spaces of connections. In Section 1.2, we review the holonomy of connections on real vector bundles, the so-called holonomy principle, and the Ambrose–Singer theorem, which relates holonomy and curvature. Section 1.3 gives a quick exposition of basic Chern–Weil representation of characteristic classes, which underlies the topological Yang–Mills energy bounds obtained in Section 1.5. In Section 1.4 we explore some variational aspects of the Yang–Mills equation, the 'Euler–Lagrange' condition for the Yang–Mills functional, and we bring in a brief discussion of gauge fixing. Finally, in Section 1.5, we recall the 4−dimensional notion of (A)SD instantons, as special first order solutions of the corresponding Yang–Mills equation, and we provide two well-known interpretations of this notion; one topological, via Chern–Weil theory, and one geometrical, in the context of complex geometry.

1.1   Connections and curvature

We will work with connections exclusively from the point of view of vector bundles. There are many good references for this topic, we will follow mostly Donaldson and Kronheimer (1990, §2.1) and Freed and Uhlenbeck (1984, §2). The last two topics, a Bochner–Weitzenböck formula and Sobolev spaces of connections, call for more specific references, which are pointed out in the text.

$G-$**bundles.**   Let $\pi\colon E \to M$ be a $\mathbb{K}-$vector bundle of rank r and structure group $G \subseteq \mathrm{GL}(r,\mathbb{K})$; henceforth, we will say simply that E is a $G-$*bundle*. This means that E admits a bundle atlas $\{(U_\alpha, \varphi_\alpha)\}$, of local trivialisations

$$\varphi_\alpha = (\pi, \phi_\alpha)\colon \pi^{-1}(U_\alpha) \to U_\alpha \times \mathbb{K}^r,$$

whose *transition functions* $\{g_{\alpha\beta}\}$ take values in G. In other words, on each nontrivial intersection $U_{\alpha\beta} := U_\alpha \cap U_\beta \neq \emptyset$, the changes in trivialisation

$$\varphi_\alpha \circ \varphi_\beta^{-1}\colon \left(U_\alpha \cap U_\beta\right) \times \mathbb{K}^r \to \left(U_\alpha \cap U_\beta\right) \times \mathbb{K}^r$$

$$(x, v) \mapsto (x, g_{\alpha\beta}(x)v),$$

define maps $g_{\alpha\beta}\colon U_{\alpha\beta} \to G$, by $g_{\alpha\beta}(x) = \phi_\alpha \circ \left(\phi_\beta|_{E_x}\right)^{-1}$. This type of atlas is also known as a $G-$*atlas* for E. A local trivialisation

$$\varphi = (\pi, \phi)\colon \pi^{-1}(U) \to U \times \mathbb{K}^r$$

is compatible with such a $G-$atlas when $\phi \circ \left(\phi_\alpha|_{E_x}\right)^{-1} \in G$, for any α with $U \cap U_\alpha \neq \emptyset$ and $x \in U \cap U_\alpha$; in this case, φ is called a $G-$*trivialisation*.

Examples of structure groups G of wide popular interest include SU(2) and SO(3). More generally, we will be interested in the groups U(r) and SU(r), when $\mathbb{K} = \mathbb{C}$, and SO(r), $r \geqslant 3$, when $\mathbb{K} = \mathbb{R}$. In any case, we will suppose G to be a *compact* Lie group, so that the Lie algebra $\mathfrak{g}$ of G admits some Ad_G-invariant inner product $\langle \cdot, \cdot \rangle_\mathfrak{g}$. In fact, in Yang–Mills theory it is common to take G to be a compact *semi-simple* Lie group, in which case there is a canonical choice of Ad_G-invariant inner product on $\mathfrak{g}$: minus the (negative definite) *Cartan–Killing form*[1] of $\mathfrak{g}$. However, because of our interest in the unitary groups U(r) (e.g. when working with Hermitian-Yang–Mills connections), in general we will not impose semi-simplicity for G. Instead, since G

---

[1] If $K_\mathfrak{g}\colon \mathfrak{g} \times \mathfrak{g} \to \mathfrak{g}$ denotes the Cartan–Killing form of $\mathfrak{g}$, i.e. the (symmetric) bilinear form on $\mathfrak{g}$ given by $K_\mathfrak{g}(X, Y) := \mathrm{tr}\,(\mathrm{ad}(X) \circ \mathrm{ad}(Y))$, for each $X, Y \in \mathfrak{g}$, then the compactness of G implies $K_\mathfrak{g}$ is a negative semi-definite bilinear form; in general, $K_\mathfrak{g}$ is non-degenerate if, and only if, $\mathfrak{g}$ is a semi-simple Lie algebra.

is compact, we assume[2] $G \subseteq O(r)$ or $G \subseteq U(r)$, according to the respective cases $\mathbb{K} = \mathbb{R}$ or $\mathbb{C}$, and fix once and for all the Ad_G–invariant inner product on $\mathfrak{g}$ to be the one induced by the canonical trace inner product:

$$\langle X, Y \rangle_\mathfrak{g} := -\mathrm{tr}(XY), \quad \forall X, Y \in \mathfrak{g}. \tag{1.1}$$

It is worth noting that minus the Cartan–Killing form of $\mathfrak{su}(r)$ (resp. $\mathfrak{so}(r)$) differs from the above choice of inner product by the constant factor $2r$ (resp. $r - 2$).

Remark 1.2. The condition $G \subseteq U(r)$ (resp. $O(r)$) implies that our bundle $E \to M$ is endowed with a *Hermitian* (resp. *Euclidean*) *metric h*, i.e. a smooth assignment of a Hermitian (resp. Euclidean) inner product h_x on E_x, for each $x \in M$. Indeed, given a trivialisation ϕ_α around $x \in U_\alpha$, define

$$h_x := \left(\phi_\alpha |_{E_x} \right)^* h_0,$$

where h_0 is the canonical Hermitian (resp. Euclidean) inner product on $\mathbb{C}^r$ (resp. $\mathbb{R}^r$). To see that this is well-defined, note that, whenever $x \in U_{\alpha\beta}$, we have $g_{\alpha\beta}(x) = \phi_\alpha \circ \left(\phi_\beta |_{E_x} \right)^{-1} \in G \subseteq U(r)$ (resp. $O(r)$), therefore

$$\begin{aligned}
\left(\phi_\beta |_{E_x} \right)^* h_0 &= \left(\phi_\beta |_{E_x} \circ \left(\phi_\alpha |_{E_x} \right)^{-1} \circ \phi_\alpha |_{E_x} \right)^* h_0 \\
&= \left(g_{\beta\alpha}(x) \circ \phi_\alpha |_{E_x} \right)^* h_0 \\
&= \left(\phi_\alpha |_{E_x} \right)^* h_0.
\end{aligned}$$

One may readily check that $h \colon x \mapsto h_x$ is a smooth assignment, i.e. for each pair of smooth local sections $s, t \in \Gamma(E|_U)$ over a neighbourhood $U \subseteq M$, the map $h(s,t) \colon x \mapsto h_x(s,t)$ is a smooth $\mathbb{K}$–valued function on U. Moreover, since each $\phi_\alpha |_{E_x}$ is a $\mathbb{K}$–linear isomorphism, it is clear that each h_x is a Hermitian (resp. Euclidean) inner product on E_x.

Conversely, if we start with a complex (resp. real) vector bundle $E \to M$ endowed with a Hermitian (resp. Euclidean) metric h, then the usual Gram–Schmidt process ensures the existence of local orthonormal frames for E, which are just another way to speak of $U(r)-$ (resp. $O(r)-$) local trivialisations for E. In particular, a $U(r)$–bundle $E \to M$ is just a complex vector bundle of rank r over M, endowed with a Hermitian metric h, also referred to as a *unitary* (or *Hermitian*) vector bundle.

---

[2]Every continuous representation $\rho \colon G \to \mathrm{Aut}(V)$ of a compact Lie group G into a finite-dimensional $\mathbb{K}$–vector space V is *unitary*, i.e. the G–module V admits a $\rho(G)$–invariant inner product.

Similarly, a SU(r)–bundle E is just a U(r)–bundle endowed with a fixed trivialisation τ on its top exterior power $\Lambda^r E^*$ (i.e., a section $\tau \in \Gamma(\Lambda^r E^*)$ assigning to each $x \in M$ an orientation $0 \neq \tau(x) \in \Lambda^r E_x^*$ on E_x); a local SU(r)–trivialisation is one for which the associated local frame is orthonormal and oriented. $\Diamond$

We denote by $\mathrm{Aut}_G(E)$ the **bundle of G–automorphisms** of E, i.e. $\mathrm{Aut}_G(E)$ is the 'bundle of groups' over M whose fibre at $x \in M$ consists of all $\mathrm{GL}(E_x)$, acting as G–isomorphisms on E_x, that is, all $g \in \mathrm{GL}(E_x)$ whose matrix representation with respect to some (and therefore any) local G–trivialisation of E lies in $G \subseteq \mathrm{GL}(r, \mathbb{K})$. The space of smooth sections of $\mathrm{Aut}_G(E)$ is denoted by $\mathcal{G}(E)$, and is called the group of **gauge transformations** of E. We note that $\mathcal{G}(E)$ is endowed with a natural group structure given by pointwise composition. Alternatively, $\mathcal{G}(E)$ is naturally identified (as a group) with the set of all G–bundle automorphisms $g \colon E \to E$ (i.e. diffeomorphisms $g \colon E \to E$ covering the identity map $\mathbb{1}_M \colon M \to M$ such that, for each $x \in M$, the restriction $g_x := g|_{E_x} \colon E_x \to E_x$ lies in $\mathrm{Aut}_G(E)_x$) with the group structure given by the composition of maps.

Another important bundle in this setting is the **adjoint bundle** $\mathfrak{g}_E$ of E, the *real* vector subbundle of $\mathrm{End}(E) = E^* \otimes E$ whose fibre at a point $x \in M$ consists of all those endomorphisms $T \colon E_x \to E_x$ whose matrix representation with respect to a G–trivialisation of E lies in the (real) Lie algebra $\mathfrak{g} \subseteq \mathfrak{gl}(r, \mathbb{K})$. Alternatively, if $\{g_{\alpha\beta} \colon U_{\alpha\beta} \to G\}$ is a family of transition functions for E, then $\mathfrak{g}_E$ is the real vector bundle determined by the transition functions

$$\mathrm{Ad}(g_{\alpha\beta}) \colon U_{\alpha\beta} \to \mathrm{GL}(\mathfrak{g}),$$

where $\mathrm{Ad} \colon G \to \mathrm{GL}(\mathfrak{g})$ is the canonical adjoint action of G on $\mathfrak{g}$.

Now, since G is a compact Lie group, the Lie algebra $\mathfrak{g}$ is *reductive*, meaning that its Levi decomposition[3] has the form

$$\mathfrak{g} = \mathfrak{s} \oplus \mathfrak{z}(\mathfrak{g}), \tag{1.3}$$

where $\mathfrak{s}$ is a semi-simple ideal and $\mathfrak{z}(\mathfrak{g})$ is the centre of $\mathfrak{g}$. (In particular, if $Z(G)$ denotes the centre of G, the compact Lie group $G/Z(G)$ is semi-simple, with Lie algebra $\mathfrak{s}$.) Furthermore, since $G \subseteq \mathrm{GL}(r, \mathbb{K})$, we have explicitly:

$$\mathfrak{s} = \mathfrak{g} \cap \mathfrak{sl}(r, \mathbb{K}) \quad \text{and} \quad \mathfrak{z}(\mathfrak{g}) = \mathfrak{g} \cap \mathfrak{l},$$

where $\mathfrak{l} \subseteq \mathfrak{gl}(r, \mathbb{K})$ denotes the Lie algebra of scalar matrices. Every element $X \in \mathfrak{g}$ decomposes accordingly as

$$X = \left(X - \frac{1}{r}\mathrm{tr}(X)\mathbb{1} \right) \oplus \left(\frac{1}{r}\mathrm{tr}(X)\mathbb{1} \right) \in \mathfrak{s} \oplus \mathfrak{z}(\mathfrak{g}).$$

---

[3] If $\mathfrak{r}(\mathfrak{g})$ denotes the (solvable) radical of a finite-dimensional (real) Lie algebra $\mathfrak{g}$, then $\mathfrak{g}$ is the semi-direct product of $\mathfrak{r}(\mathfrak{g})$ and a (necessarily semi-simple) subalgebra $\mathfrak{s}$.

The $\mathfrak{s}$–component (or *trace-free component*) of X will be denoted by X^0.

It follows from decomposition (1.3) that the adjoint bundle $\mathfrak{g}_E$ splits as

$$\mathfrak{g}_E = \mathfrak{g}_E^{(0)} \oplus \mathfrak{z}(\mathfrak{g}), \tag{1.4}$$

where $\mathfrak{g}_E^{(0)}$ is the adjoint bundle of $E \times_G G/Z(G)$, consisting of trace-free endomorphisms in $\mathfrak{g}_E$, and $\mathfrak{z}(\mathfrak{g})$ is the trivial bundle with fibre $\mathfrak{z}(\mathfrak{g})$.

Connections.   We now recall the definition of a connection on E from the covariant derivative (Koszul) point of view.

Definition 1.5. A smooth **connection** (or *covariant derivative*) ∇ on E is a $\mathbb{K}$–linear map

$$\nabla \colon \Gamma(E) \to \Gamma(T^*M \otimes E)$$

satisfying the *Leibniz rule*

$$\nabla(fs) = \mathrm{d}f \otimes s + f\nabla s, \quad \text{for each } f \in C^\infty(M), \text{ and } s \in \Gamma(E). \tag{1.6}$$

Remark 1.7. When E is a *complex* vector bundle ($\mathbb{K} = \mathbb{C}$), its space of sections $\Gamma(E)$ has a natural $C^\infty(M, \mathbb{C})$–module structure. This induces a natural $C^\infty(M, \mathbb{C})$–module structure on $\Gamma(T^*M \otimes E)$, which canonically identifies it with $\Gamma(T^*M_\mathbb{C} \otimes_{C^\infty(M,\mathbb{C})} E)$. For instance, when (M, J) is an almost complex manifold, one may think of ∇ as an operator on $\Gamma(E)$ taking values in $\Gamma(T^*M_\mathbb{C} \otimes_{C^\infty(M,\mathbb{C})} E)$, instead of $\Gamma(T^*M \otimes E)$ (we will do so in §1.5). In this case, it makes sense to write $\mathrm{d}f \otimes s = \mathrm{d}f_1 \otimes s + \mathrm{d}f_2 \otimes (\mathbf{i}s)$ for each $f = f_1 + \mathbf{i}f_2 \in C^\infty(M, \mathbb{C})$, with $f_i \in C^\infty(M)$, so that the Leibniz rule (1.6) holds more generally for *complex*-valued smooth functions. $\diamondsuit$

In what follows we list some important properties of connections:

(i) *The difference of two connections is a tensor.* It follows from (1.6) that $\nabla - \nabla'$ is an *algebraic operator* (i.e. $C^\infty(M)$–linear), hence it defines an element $A \in \Omega^1(M, \mathrm{End}(E))$ such that

$$\left(\nabla - \nabla'\right)s = As, \quad \text{for each } s \in \Gamma(E) \equiv \Omega^0(M, E).$$

Here A acts algebraically on sections, by contraction: [4]

$$\Omega^0(M, E) \times \Omega^1(M, \mathrm{End}(E)) \to \Omega^1(M, E).$$

---

[4]e.g. $(s, \omega \otimes T) \mapsto \omega \otimes (Ts)$, for $\omega \in \Omega^1(M)$, $T \in \Gamma(\mathrm{End}(E))$ and $s \in \Omega^0(M, E)$.

Conversely, given a connection ∇ on E and $A \in \Omega^1(M, \mathrm{End}(E))$, one immediately verifies linearity and the Leibniz rule for

$$\nabla' := \nabla + A \colon \Omega^0(M, E) \to \Omega^1(M, E),$$

hence it defines another connection on E. Therefore, the space of connections on $E \to M$ is an affine space[5] modeled on $\Omega^1(M, \mathrm{End}(E))$.

(ii) *Connections are local operators.* Another consequence of Definition 1.5 is that a connection ∇ is a *local operator*, in the sense that it *decreases support*: if $s \in \Gamma(E)$ vanishes on some open subset $U \subseteq M$, then ∇s also vanishes on U. By linearity, this amounts to saying that the value of ∇s at x depends only on the values of s near x. Indeed, let $x \in U$ and pick a cutoff function $\rho \in C^\infty(M)$ supported on U and equal to 1 in a neighbourhood $V \Subset U$ of x, i.e. $\mathrm{supp}(\rho) \subseteq U$ and $\rho|_V \equiv 1$, for some open $V \ni x$, $\overline{V} \subset U$. Then, $\rho s \equiv 0$ and, by linearity,

$$\nabla(\rho s) \equiv 0.$$

On the other hand, as $s(x) = 0$ and $\rho(x) = 1$, by the Leibniz rule,

$$\begin{aligned}
\nabla(\rho s)(x) &= (\mathrm{d}\rho)(x) \otimes s(x) + \rho(x)(\nabla s)(x) \\
&= (\nabla s)(x).
\end{aligned}$$

Therefore $(\nabla s)(x) = 0$, as claimed.

In particular, if ∇ is a connection on E, then it restricts to a connection on $E|_U$ over any open set $U \subseteq M$. Thus if $\{U_\alpha\}$ is an open cover of M, then ∇ is completely determined by the induced connections ∇_α on each of the restrictions $E|_{U_\alpha}$.

(iii) *Connections are covariant.* Connections can be pulled back by a smooth map $f \colon M' \to M$. If $\{g_{\alpha\beta}\}$ is a family of transition functions for a $G-$bundle E, subordinated to an open cover $\{U_\alpha\}$ of M, then the induced bundle $f^*E \to M'$ is also a $G-$bundle, determined by the family $\{f^*g_{\alpha\beta}\}$, subordinated to the cover $\{f^{-1}(U_\alpha)\}$ of M', with total space

$$f^*E = \{(x', v) \in M' \times E : f(x') = \pi(v)\}.$$

Each local section $s \in \Gamma(E|_U)$ induces $f^*s \in \Gamma(f^*E|_{f^{-1}(U)})$, defined by

$$(f^*s)(x') := s(f(x')), \quad \forall x' \in f^{-1}(U).$$

---

[5] The existence of connections on a vector bundle over a *paracompact* manifold follows by a standard application of partitions of unity, so we adopt that assumption henceforth.

If $\{e_1, \ldots, e_r\}$ is a local frame of E over U, it is quite easy to see that $\{f^*e_1, \ldots, f^*e_r\}$ is a local frame for f^*E over $f^{-1}(U)$.

Given a connection ∇ on E and a G−atlas $\{(U_\alpha, \varphi_\alpha)\}$ for E, the *pullback connection* $f^*\nabla$ on f^*E is defined, in each induced local frame $\{f^*e_i^\alpha : i = 1, \ldots, r\}$, by

$$(f^*\nabla)(f^*e_i^\alpha) := f^*(\nabla e_i^\alpha), \tag{1.8}$$

where in the RHS of the last equation f^* acts as the natural extension of $f^*(\omega \otimes s) := (f^*\omega) \otimes (f^*s)$, for $\omega \in \Omega^1(U_\alpha)$ and $s \in \Gamma(E|_{U_\alpha})$. This means that, if $\{A_\alpha\}$ is the collection of 'gauge potentials' associated to ∇, then $\{f^*A_\alpha\}$ is the collection of 'gauge potentials' associated to $f^*\nabla$ on the induced local trivialisations for f^*E; see the next paragraph.

Local description of connections. Combining properties (i) and (ii) above, we get the following local description of connections. Consider an atlas $\{(U_\alpha, \varphi_\alpha)\}$ of local trivialisations for E. Then, we may write

$$\nabla_\alpha = d + A_\alpha, \tag{1.9}$$

where d is the *trivial product connection* on $U_\alpha \times \mathbb{K}^r$, which takes a section $s = (s_1, \ldots, s_r)$ to[6] $ds = (ds_1, \ldots, ds_r)$, and[7] $A_\alpha \in \Omega^1(U_\alpha, \mathfrak{gl}(r, \mathbb{K}))$. The meaning of the above equality is that, identifying (via φ_α) local sections of $E|_{U_\alpha}$ with (column) vector-valued functions, the induced covariant derivative on $E|_{U_\alpha}$ acts on sections as the sum $d + A_\alpha$. The matrix A_α of local 1−forms is called the *connection matrix* or *gauge potential* of ∇ with respect to $(U_\alpha, \varphi_\alpha)$.

In an overlap $U_\alpha \cap U_\beta \neq \emptyset$, a straightforward computation shows that the gauge potentials A_α and A_β are related by

$$A_\alpha = g_{\alpha\beta} A_\beta g_{\alpha\beta}^{-1} + g_{\alpha\beta} dg_{\alpha\beta}^{-1}. \tag{1.10}$$

G−**connections.** A connection ∇ on E is called a G−**connection** if its associated gauge potentials A_α with respect to local G−trivializations $(U_\alpha, \varphi_\alpha)$ of E lie in $\Omega^1(U_\alpha, \mathfrak{g})$. For example, if $G = \mathrm{U}(r)$ and $\langle \cdot, \cdot \rangle$ is the associated metric on E, then the condition for ∇ to be a G−connection may be rephrased globally as:

$$d\langle s_1, s_2 \rangle = \langle \nabla s_1, s_2 \rangle + \langle s_1, \nabla s_2 \rangle, \quad \forall s_1, s_2 \in \Gamma(E).$$

---

[6]we also denote by d the exterior derivative operator.
[7]We think of $\mathfrak{gl}(r, \mathbb{K})$ as the trivial bundle $U_\alpha \times \mathfrak{gl}(r, \mathbb{K})$ over U_α.

Here $\langle \cdot, \cdot \rangle$ is naturally extended so that

$$\langle \omega \otimes s, t \rangle = \langle s, \omega \otimes t \rangle = \langle s, t \rangle \omega,$$

whenever $\omega \in \Omega^1(M)$ and $s, t \in \Gamma(E)$.

We see that G−connections differ by an element of $\Omega^1(M, \mathfrak{g}_E)$ rather than just $\Omega^1(M, \mathrm{End}(E))$, so that the **space of smooth G−connections** on E, hereafter denoted by $\mathcal{A}(E)$, is an (infinite-dimensional) affine space modeled on $\Omega^1(M, \mathfrak{g}_E)$. Thus, when we fix a smooth reference G−connection ∇_0,

$$\mathcal{A}(E) = \{\nabla_0 + A : A \in \Omega^1(M, \mathfrak{g}_E)\}.$$

We use this affine structure to endow $\mathcal{A}(E)$ with the C^∞_{loc}−topology coming from the model $\Omega^1(M, \mathfrak{g}_E)$. By definition, a sequence $\{\nabla_i\} \subseteq \mathcal{A}(E)$ converges to $\nabla \in \mathcal{A}(E)$ if, and only if, $\{\nabla_i - \nabla\} \subseteq \Omega^1(M, \mathfrak{g}_E)$ converges to zero in C^∞_{loc}−topology[8] on M.

Convention 1.11. Unless otherwise stated, from now on we drop the prefix G− and assume we are dealing only with G−objects. For instance, 'a connection on E' will actually mean 'a G−connection on E', a 'local trivialisation for E' will actually mean a 'a local G−trivialisation of E', and so on.

Now we turn attention to some important differential operators induced by a connection $\nabla \in \mathcal{A}(E)$. We have from the outset the collection of *covariant exterior derivatives*

$$\mathrm{d}_\nabla \colon \Omega^k(M, E) \to \Omega^{k+1}(M, E), \quad k \geqslant 0,$$

uniquely determined by the following properties (see Madsen and Tornehave (1997, p. 170, Lemma 17.6)):

(i) d_∇ is $\mathbb{K}$−linear, for each $k \geqslant 0$;

(i) $\mathrm{d}_\nabla = \nabla$ on $\Omega^0(M, E)$;

(ii) $\mathrm{d}_\nabla(\omega \wedge \xi) = \mathrm{d}\omega \wedge \xi + (-1)^k \omega \wedge \mathrm{d}_\nabla \xi$, for each $\omega \in \Omega^k(M)$ and $\xi \in \Omega^l(M, E)$.

Here $\wedge \colon \Omega^k(M) \times \Omega^l(M, E) \to \Omega^{k+l}(M, E)$ is the naturally extended wedge product acting trivially on the E−component.

---

[8]See the first paragraph of Section A.6 for a construction of such topology in a simplified context.

The curvature of a connection. Moreover, the composition

$$d_\nabla \circ d_\nabla \colon \Omega^0(M, E) \to \Omega^2(M, E)$$

is $C^\infty(M)-$linear: indeed, for $f \in C^\infty(M)$ and $s \in \Gamma(E)$ we have

$$
\begin{aligned}
d_\nabla \circ d_\nabla(fs) &= d_\nabla(df \otimes s + f\nabla s) \\
&= d^2 f \otimes s - df \wedge \nabla s + df \wedge \nabla s + f\, d_\nabla \circ d_\nabla s \\
&= f\, d_\nabla \circ d_\nabla s.
\end{aligned}
$$

Hence, there exists a unique section $F_\nabla \in \Omega^2(M, \mathrm{End}(E))$, called the **curvature** of ∇, such that

$$F_\nabla s = (d_\nabla \circ d_\nabla)s, \quad \forall s \in \Gamma(E).$$

Local description of the curvature. Consider an atlas of local trivialisations $(U_\alpha, \varphi_\alpha)$ for E. Let again A_α be the associated gauge potentials of ∇ and let $F_\alpha := [F_\nabla]_\alpha$ be the local matrix representations of F_∇. Then, a local computation gives the following *Cartan formula*:

$$F_\alpha = dA_\alpha + A_\alpha \wedge A_\alpha, \tag{1.12}$$

where $A_\alpha \wedge A_\alpha$ is the matrix of local $2-$forms

$$(A_\alpha \wedge A_\alpha)^i_j := \sum_k (A_\alpha)^i_k \wedge (A_\alpha)^k_j, \quad 1 \leqslant i, j \leqslant r.$$

Here, ξ^i_j denotes the local components of some element $\xi \in \Gamma(\mathrm{End}(E))$, with respect to the local frame induced by φ_α on $\mathrm{End}(E)$; more intrinsically, there is a natural extension of the wedge product to

$$\Omega^\bullet(M, \mathrm{End}(E)) = \bigoplus_{k \geqslant 0} \Omega^k(M, \mathrm{End}(E))$$

such that

$$(\omega \otimes T) \wedge (\eta \otimes S) = (\omega \wedge \eta) \otimes (T \circ S),$$

for each $\omega, \eta \in \Omega^\bullet(M)$ and $S, T \in \Gamma(\mathrm{End}(E))$.

Using (1.12) and (1.10), one further shows that

$$F_\alpha = g_{\alpha\beta} F_\beta g_{\alpha\beta}^{-1} = \mathrm{Ad}(g_{\alpha\beta})F_\beta, \quad \text{on } U_{\alpha\beta} \neq \emptyset. \tag{1.13}$$

Shrinking U_α if necessary, and considering local coordinates $(x^1, \ldots, x^n)$ and write

$$A_\alpha = A_{\alpha,i} \otimes \mathrm{d}x^i, \quad \text{for } A_{\alpha,i} \in \mathfrak{g}, \tag{1.14}$$

and

$$F_\alpha = \frac{1}{2} F_{\alpha,ij} \otimes \mathrm{d}x^i \wedge \mathrm{d}x^j, \quad \text{for } F_{\alpha,ij} \in \mathfrak{gl}(r, \mathbb{K}). \tag{1.15}$$

It then follows from (1.12) that

$$F_{\alpha,ij} = \partial_i A_{\alpha,j} - \partial_j A_{\alpha,i} + [A_{\alpha,i}, A_{\alpha,j}], \tag{1.16}$$

where $[\cdot, \cdot]$ is the commutator of $\mathfrak{g} \subseteq \mathfrak{gl}(r, \mathbb{K})$. In particular, we have $F_{\alpha,ij} \in \mathfrak{g}$ for each $i, j = 1, \ldots, n$. Thus, the curvature F_∇ lies actually in $\Omega^2(M, \mathfrak{g}_E)$.

Furthermore, ∇ also induces a connection on the bundle $\mathrm{End}(E)$, still denoted by ∇, which acts on $T \in \Gamma(\mathrm{End}(E))$ by the tautological Leibniz's rule

$$(\nabla T)(s) := \nabla(Ts) - T(\nabla s), \quad \text{for each } s \in \Gamma(E), \tag{1.17}$$

where $T(\nabla s)$ denotes the action of the endomorphism T on the E component of ∇s. This connection in fact reduces to a connection on $\mathfrak{g}_E \subseteq \mathrm{End}(E)$, since ∇ is a $G-$connection. As before, this induces operators

$$\mathrm{d}_\nabla \colon \Omega^k(M, \mathfrak{g}_E) \to \Omega^{k+1}(M, \mathfrak{g}_E), \quad k \geqslant 0.$$

If $\xi \in \Omega^p(M, \mathfrak{g}_E)$ and $\xi_\alpha := [\xi]_\alpha$ is the local representation via a local trivialisation $(U_\alpha, \varphi_\alpha)$, then one can show that

$$[\mathrm{d}_\nabla \xi]_\alpha = \mathrm{d}\xi_\alpha + [A_\alpha, \xi_\alpha], \tag{1.18}$$

where $[\omega \otimes T, \eta \otimes S] := (\omega \wedge \eta) \otimes [T, S]_\mathfrak{g}$ is the *graded commutator*; more generally, if $\eta \in \Omega^q(M, \mathfrak{g}_E)$,

$$[\xi, \eta] := \xi \wedge \eta - (-1)^{pq} \eta \wedge \xi. \tag{1.19}$$

Lemma 1.20 (Bianchi identity). *A smooth connection* $\nabla \in \mathcal{A}(E)$ *satisfies*

$$\mathrm{d}_\nabla F_\nabla = 0. \tag{1.21}$$

Proof. It suffices to check the identity in a local trivialisation. By (1.18) and Cartan's formula (1.12), we have

$$\begin{aligned}
[\mathrm{d}_\nabla F_\nabla]_\alpha &= \mathrm{d}F_\alpha + [A_\alpha, F_\alpha] \\
&= \mathrm{d}(\mathrm{d}A_\alpha + A_\alpha \wedge A_\alpha) + [A_\alpha, \mathrm{d}A_\alpha + A_\alpha \wedge A_\alpha].
\end{aligned}$$

Now, by the Leibniz rule and (1.19),

$$\mathrm{d}(\mathrm{d}A_\alpha + A_\alpha \wedge A_\alpha) = \mathrm{d}A_\alpha \wedge A_\alpha - A_\alpha \wedge \mathrm{d}A_\alpha$$

and

$$[A_\alpha, \mathrm{d}A_\alpha + A_\alpha \wedge A_\alpha] = A_\alpha \wedge \mathrm{d}A_\alpha - \mathrm{d}A_\alpha \wedge A_\alpha.$$

Summing these equations we get the desired result. $\qquad\square$

Gauge equivalence. Let us recall the concept of gauge equivalence for connections on E, in terms of the canonical action of $\mathcal{G}(E)$ on $\Gamma(E)$: a gauge transformation $g \in \mathcal{G}(E)$ acts on a section $s \in \Gamma(E)$, giving rise to the new section $gs \in \Gamma(E)$ defined by

$$(gs)(x) := g_x(s(x)), \quad \forall x \in M.$$

This extends to an action of $\mathcal{G}(E)$ on $\Omega^k(M, E)$ by acting trivially on the form part. We can define the following 'pullback' action[9] of $\mathcal{G}(E)$ on the space of smooth $G-$connections $\mathcal{A}(E)$: an element $g \in \mathcal{G}(E)$ acts on $\nabla \in \mathcal{A}(E)$ by

$$g^*\nabla := g^{-1} \circ \nabla \circ g,$$

i.e. $g^*\nabla \colon \Omega^0(M, E) \to \Omega^1(M, E)$ is the map given by

$$(g^*\nabla)(s) := g^{-1}\nabla(gs), \quad \forall s \in \Omega^0(M, E).$$

This defines indeed a $G-$connection on E. First of all, $g^*\nabla$ is clearly a $\mathbb{K}-$linear map. To check the Leibniz rule, let $f \in C^\infty(M)$ and $s \in \Omega^0(M, E)$; since the actions of $C^\infty(M)$ and $\mathcal{G}(E)$ on $\Omega^k(M, E)$ commutes, we have

$$\begin{aligned}
(g^*\nabla)(fs) &= g^{-1}(\nabla g(fs)) = g^{-1}(\nabla f(gs)) \\
&= g^{-1}(\mathrm{d}f \otimes (gs) + f\nabla(gs)) = \mathrm{d}f \otimes s + f(g^{-1}\nabla(gs)) \\
&= \mathrm{d}f \otimes s + f(g^*\nabla)s.
\end{aligned}$$

Moreover, if (U, φ) is a local $(G-)$trivialisation for E and we let A and g^*A be the respective gauge potentials of ∇ and $g^*\nabla$, then we the following transformation law is easy to compute:

$$g^*A = g^{-1}Ag + g^{-1}\mathrm{d}g = \mathrm{Ad}(g^{-1})A + g^*\theta_{MC}, \qquad (1.22)$$

---

[9]Here, the term 'pullback' alludes to the fact this is a *right*-action. Some authors defines the action of $\mathcal{G}(E)$ on $\mathcal{A}(E)$ to be the corresponding 'pushforward' left-action: $g \cdot \nabla := g \circ \nabla \circ g^{-1}$.

where $g, g^{-1} \colon U \to G$ are seen here as sections of $U \times G \simeq \mathrm{Aut}(E|_U)$ via φ, and θ_{MC} is the Maurer–Cartan form[10] of G. This shows that $g^* \nabla$ is in fact a G−connection. Finally, one checks directly from the definition that $g^*(h^*\nabla) = (h \circ g)^*\nabla$ for each $g \in \mathcal{G}(E)$ and $\nabla \in \mathcal{A}(E)$, characterizing a right action.

We say that connections $\nabla, \nabla' \in \mathcal{A}(E)$ are **gauge equivalent** if they lie in the same $\mathcal{G}(E)$-orbit, i.e. if there exists $g \in \mathcal{G}(E)$ such that $\nabla' = g^*\nabla$. Clearly,

$$F_{g^*\nabla} = g^{-1} F_\nabla g, \quad \forall g \in \mathcal{G}(E), \nabla \in \mathcal{A}(E). \tag{1.23}$$

Laplacians induced by connections. We now introduce some other important differential operators induced by connections on *real* vector bundles defined over (oriented) *Riemannian* manifolds.

Let (M, g) be an oriented Riemannian manifold, and let $F \to M$ be a real vector bundle endowed with a fibre metric $\langle \cdot, \cdot \rangle$. Recall that g distinguishes a *Levi-Civita* connection, the unique torsion-free $\mathrm{O}(n)$−connection D^g on TM. Tensoring with D^g, a connection $\nabla \in \mathcal{A}(F)$ on $F \to M$ induces connections

$$\nabla \colon \Omega^k(M, F) \to \Gamma(T^*M \otimes \Lambda^k T^*M \otimes F), \quad \text{for each } k \geqslant 0. \tag{1.24}$$

Recalling the definition of the covariant exterior differential operator d_∇ on $\Omega^k(M, F)$, we see that $\mathrm{d}_\nabla = \wedge \circ \nabla$, under the natural map $\wedge \colon \Lambda^1 \otimes \Lambda^k \to \Lambda^{k+1}$.

The metric g naturally induces metrics on every tensor bundle of M. In particular, we get (Euclidean) metrics on every exterior power $\Lambda^k T^*M$. Tensoring with the metric $\langle \cdot, \cdot \rangle$ on F gives rise to (Euclidean) metrics, still denoted by $\langle \cdot, \cdot \rangle$, on the bundles $\Lambda^k T^*M \otimes F$. One readily checks that the induced connections defined in the above paragraph are compatible with the respective induced metrics.

Let $\mathrm{d}V_g$ be the Riemannian volume n−form on (M, g) determined by the orientation. The Hodge star operator

$$* \colon \Lambda^k T^*M \to \Lambda^{n-k} T^*M$$

isomorphically interchanges forms of complementary degree by the relation $\alpha \wedge *\beta = (\alpha, \beta)_g \mathrm{d}V_g$, where $\alpha, \beta \in \Lambda^k T^*M$ and $(\cdot, \cdot)_g$ denotes the induced metric on $\Lambda^k T^*M$. More generally, given any vector bundle $W \to M$, we define $* \colon \Omega^k(M, W) \to \Omega^{n-k}(M, W)$ by acting trivially on the W part: $*(\alpha \otimes s) := (*\alpha) \otimes s$.

---

[10] $\theta_{MC} \in \Omega^1(G, \mathfrak{g})$ is the unique $\mathfrak{g}$-valued left-invariant 1−form on G such that $(\theta_{MC})_1 \colon \mathfrak{g} \to \mathfrak{g}$ is the identity map.

If $\xi, \eta \in \Omega^k(M, F)$, at least one of which has compact support, we define their L^2-**inner product** by

$$\langle \xi, \eta \rangle_{L^2} := \int_M \langle \xi, \eta \rangle \mathrm{d}V_g.$$

This gives rise to formal L^2-*adjoint* operators for ∇ and d_∇:

$$\nabla^*: \Gamma(T^*M \otimes \Lambda^k T^*M \otimes F) \to \Omega^k(M, F),$$
$$\mathrm{d}_\nabla^*: \Omega^{k+1}(M, E) \to \Omega^k(M, E).$$

For example, d_∇^* is characterised by the equation

$$\langle \mathrm{d}_\nabla \xi, \eta \rangle_{L^2} = \langle \xi, \mathrm{d}_\nabla^* \eta \rangle_{L^2},$$

which is valid for forms ξ, η at least one of which has compact support. Furthermore, using Stokes' theorem, one can show that

$$\mathrm{d}_\nabla^* = (-1)^{n(k+1)+1} * \mathrm{d}_\nabla *, \quad \text{on } \Omega^k(M, F).$$

The above notions naturally induce two important second order operators acting on $\Omega^k(M, F)$:

- the **generalised Hodge–de Rham Laplacian**

$$\Delta_\nabla := \mathrm{d}_\nabla \mathrm{d}_\nabla^* + \mathrm{d}_\nabla^* \mathrm{d}_\nabla : \Omega^k(M, F) \to \Omega^k(M, F),$$

- the **covariant** (or *rough*) **Laplacian**

$$\nabla^* \nabla : \Omega^k(M, F) \to \Omega^k(M, F).$$

In terms of an orthonormal local frame $(e_1, \ldots, e_n)$ of TM, we have

$$\nabla^* \nabla \xi = -\sum_{j=1}^n \nabla^2(e_j, e_j)\xi,$$

where $\nabla^2(X, Y) := \nabla_X \nabla_Y - \nabla_{D_X^g Y}$ is the invariantly defined *Hessian operator*.

A Bochner–Weitzenböck Formula.　　(cf. Bourguignon and Lawson Jr (1981, pp. 199-200)) Consider now our $\mathbb{K}$−vector bundle $E \to M$ with compact structure group G, and fix a smooth connection $\nabla \in \mathcal{A}(E)$. By Ad−invariance, the inner product $\langle \cdot, \cdot \rangle_{\mathfrak{g}}$ naturally induces a metric [11] on the real vector bundle $\mathfrak{g}_E$. Then, letting $F = \mathfrak{g}_E$ in the discussion of the previous paragraph, and considering the induced connection, still denoted by ∇, on $\mathfrak{g}_E$ (see 1.17), the corresponding operators Δ_∇ and $\nabla^*\nabla$ act on $\mathfrak{g}_E$−valued k−forms.

These operators have the same principal symbol and their difference is a zero-order (algebraic) operator, i.e. it is $C^\infty(M)$−linear. The precise difference between these operators on the space $\Omega^2(M, \mathfrak{g}_E)$ is identified by a **Bochner–Weitzenböck formula**, which we will now state.

Fix an orthonormal local frame $\{e_1, \ldots, e_n\}$ of TM. Recall that the *Ricci transformation* $\mathrm{Ric}^g : T_x M \to T_x M$ is given by

$$\mathrm{Ric}^g(X) = \sum_{j=1}^{n} R^g(X, e_j)e_j,$$

where R^g stands for the *Riemann curvature tensor* of g. We extend the Ricci transformation to act on 2−forms by

$$(\mathrm{Ric}^g \wedge I)(X, Y) := \mathrm{Ric}^g(X) \wedge Y + X \wedge \mathrm{Ric}^g(Y), \quad \forall X, Y \in \mathfrak{X}(M).$$

Now define a zero-order operator $\mathscr{F}_\nabla : \Omega^2(M, \mathfrak{g}_E) \to \Omega^2(M, \mathfrak{g}_E)$ by

$$\mathscr{F}_\nabla(\xi)(X, Y) := \sum_{j=1}^{n} \left\{ [F_\nabla(e_j, X), \xi(e_j, Y)] - [F_\nabla(e_j, Y), \xi(e_j, X)] \right\},$$

for all $X, Y \in \mathfrak{X}(M)$. For each $\xi \in \Omega^2(M, \mathfrak{g}_E)$, we write

$$(\xi \circ \mathrm{Ric}^g \wedge I)(X, Y) := \xi(\mathrm{Ric}^g(X), Y) + \xi(X, \mathrm{Ric}^g(Y)), \quad \text{and}$$

$$(\xi \circ 2R^g)(X, Y) := \sum_{j=1}^{n} \xi(e_j, R^g(X, Y)e_j), \quad \forall X, Y \in \mathfrak{X}(M).$$

The following formula can be found in Bourguignon and Lawson Jr (ibid., Theorem 3.10, p. 200).

---

[11] In a local trivialisation, elements of $\mathfrak{g}_E$ are represented by matrices in $\mathfrak{g}$. Moreover, any two such representations differ by the adjoint action of G on $\mathfrak{g}$. Thus there is well-defined induced metric $\langle \cdot, \cdot \rangle$ on $\mathfrak{g}_E$ such that, for all $T, S \in \Omega^0(\mathfrak{g}_E)$, we have locally

$$\langle T, S \rangle|_{U_\alpha} = \langle [T]_\alpha, [S]_\alpha \rangle_{\mathfrak{g}}.$$

Theorem 1.25 (Bochner–Weitzenböck formula).
For any $\xi \in \Omega^2(M, \mathfrak{g}_E)$, we have

$$\Delta_\nabla \xi = \nabla^* \nabla \xi + \xi \circ (\mathrm{Ric}^g \wedge I + 2R^g) + \mathscr{F}_\nabla(\xi).$$

Sobolev spaces of connections. In this paragraph, we suppose M is a compact manifold. Let $E \to M$ be a $G-$bundle. Given $k \in \mathbb{N}$ and $1 \leqslant p < \infty$, we want to introduce the Sobolev space $\mathcal{A}^{k,p}(E)$ of $W^{k,p}$ connections on E. Our main reference for this topic is Wehrheim (2004, Appendix A and Appendix B))

The metric on the adjoint bundle $\mathfrak{g}_E$ (determined by the Ad_G-invariant inner product on $\mathfrak{g}$), combined with the metric induced by g on T^*M, gives rise to a natural (tensor product) metric on the bundle $T^*M \otimes \mathfrak{g}_E$. If we fix a smooth connection $\nabla_0 \in \mathcal{A}(E)$, this induces (twisting by the Levi-Civita connection) a compatible connection on $T^*M \otimes \mathfrak{g}_E$. Thus, we can speak of the Sobolev spaces $W^{k,p}(M, T^*M \otimes \mathfrak{g}_E)$, for each $1 \leqslant p < \infty$ and $k \in \mathbb{N}_0$ (cf. Section B.2 of Appendix B). In this context, we can define the **Sobolev space of $W^{k,p}$ connections** on E by

$$\mathcal{A}^{k,p}(E) := \{\nabla_0 + A : A \in W^{k,p}(M, T^*M \otimes \mathfrak{g}_E)\}.$$

Since M is compact, we know from Theorem B.12 that $W^{k,p}(M, T^*M \otimes \mathfrak{g}_E)$ does not depend on the choices of metrics and compatible connections on the involved bundles. Moreover, since any two smooth reference connections $\nabla_0, \nabla_0' \in \mathcal{A}(E)$ differ by an element of $\Omega^1(M, \mathfrak{g}_E)$ and, by compactness of M, there is a bounded inclusion $\Omega^1(M, \mathfrak{g}_E) \hookrightarrow W^{k,p}(M, T^*M \otimes \mathfrak{g}_E)$, we see that $\mathcal{A}^{k,p}(E)$ is well-defined.

We topologize $\mathcal{A}^{k,p}(E)$ using its affine structure: by definition, a sequence $\{\nabla_i\} \subseteq \mathcal{A}^{k,p}(E)$ converges to $\nabla \in \mathcal{A}^{k,p}(E)$ if, and only if, $\|\nabla_i - \nabla\|_{p,k} \to 0$ as $i \to \infty$.

We know that a smooth gauge transformation $g \in \mathcal{G}(E)$ acts on a smooth connection $\nabla = \nabla_0 + A \in \mathcal{A}(E)$ by

$$g^*\nabla = g^{-1} \circ \nabla \circ g = \nabla_0 + g^{-1}\nabla_0 g + g^{-1}Ag.$$

Naturally enough, the relevant group of gauge transformations in the context of $W^{k,p}-$connections is [12]

$$\mathcal{G}^{k+1,p}(E) := W^{k+1,p}(M, \mathrm{Aut}(E)).$$

In fact, using the Sobolev embedding theorem, one can prove the following Wehrheim (ibid., Lemma A.5, p. 175 & Lemma A.6, p. 176):

---

[12] The heuristic is that we need one more order of regularity on g in order to $g^*A = g^{-1}\nabla_0 g + g^{-1}Ag$ lie in $W^{k,p}$ whenever $A \in W^{k,p}$.

Proposition 1.26. *For $k \in \mathbb{N}_0$ and $1 \leqslant p < \infty$ such that $(k+1)p > n$, the inclusion $\mathcal{G}^{k+1,p}(E) \subseteq C^0(M, \mathrm{Aut}_G(E))$ makes $\mathcal{G}^{k+1,p}(E)$ a topological group with respect to composition. Moreover, the pullback action $\mathcal{G}^{k+1,p}(E) \times \mathcal{A}^{k,p}(E) \to \mathcal{A}^{k,p}(E)$ is a continuous map. In particular, for $p > \frac{n}{2}$, $\mathcal{G}^{2,p}(E)$ acts continuously in $\mathcal{A}^{1,p}(E)$.*

The curvature (or *field strength*) of a smooth connection $\nabla = \nabla_0 + A \in \mathcal{A}(E)$ is

$$F_\nabla = \nabla^2 = F_{\nabla_0} + \mathrm{d}_{\nabla_0} A + [A, A] \in \Omega^2(M, \mathfrak{g}_E).$$

More generally, we have Uhlenbeck (1982a, Lemma 1.1):

Lemma 1.27. *Let $1 < p < \infty$ be such that $2p \geqslant n$. Then, the curvature map $\nabla \mapsto F_\nabla$ on $\mathcal{A}(E)$ extends to a quadratic map*

$$\mathcal{A}^{1,p}(E) \to L^p(M, \Lambda^2 T^*M \otimes \mathfrak{g}_E).$$

Sketch of proof. We know that $F_\nabla = F_{\nabla_0} + \nabla_0 A + [A, A]$, with $F_{\nabla_0} \in C^\infty$ and $\nabla_0 A \in L^p$. By the Sobolev embedding (Theorem B.13), we have $W^{1,p} \subseteq L^q$ for $\frac{1}{q} \geqslant \frac{1}{p} - \frac{1}{n}$; in this case, by Hölder's inequality, the quadratic term $A \mapsto [A, A]$ lies in $L^{q/2}$. In order to obtain $L^{q/2} \subseteq L^p$, the Sobolev embedding requires $\frac{1}{p} \geqslant \frac{2}{q} \geqslant \frac{2}{p} - \frac{2}{n}$, i.e. $2p \geqslant n$. $\qquad\square$

1.2   Holonomy groups

Fix $E \to M$ a real vector bundle of rank r endowed with a smooth connection ∇. We now briefly review the basics about holonomy groups, fixing terminology and notation which will be used in Chapter 2. The main references for this section are Joyce (2006, 2007) and Clarke and Santoro (2012).

Parallel transport.   Let $\gamma: [0,1] \to M$ be a smooth path from $x = \gamma(0)$ to $y = \gamma(1)$. A section $s \in \Gamma(E)$ is said to be ∇–**parallel along** γ when the composition $s \circ \gamma \in \Gamma(\gamma^*E)$ is $\gamma^*\nabla$–parallel [13], i.e. when

$$\left(\gamma^*\nabla\right)(s \circ \gamma) = 0. \tag{1.28}$$

Since $[0,1]$ is contractible, the induced bundle $\gamma^*E \to [0,1]$ is trivial, i.e. it admits a global frame $\{E_i\}$. Writing $s \circ \gamma = x^j E_j$ and denoting by $A = (A^i_j)$

---

[13] This pull-back notation is not entirely rigorous since $[0,1]$ is a manifold with boundary. Now, the smoothness of γ means that there exists a smooth path $\widetilde{\gamma}: \,]-\varepsilon, 1+\varepsilon[\to M$, for some $\varepsilon > 0$, such that $\widetilde{\gamma}|_{[0,1]} = \gamma$. So when we talk about $\gamma^*\nabla$ we mean in fact the restriction $\widetilde{\gamma}^*\nabla|_{[0,1]}$ (one can show this is independent of the extension $\widetilde{\gamma}$).

the gauge potential of $\gamma^*\nabla$ with respect to $\{E_i\}$, equation (1.28) translates into the linear ODE:

$$\dot{x} + Ax = 0,$$

where $x = (x^1, \ldots, x^n) \colon [0, 1] \to \mathbb{R}^r$. Thus, invoking the well-known existence and uniqueness theorem for ODE's, given an initial incidence condition $v \in E_x$, there exists a unique ∇–parallel section $s_{\gamma,v}$ along γ satisfying $s_{\gamma,v}(x) = v$. Moreover, by linearity of the equation, the solution depends linearly on the initial condition. This allows us to define the linear map

$$P_\gamma \colon E_x \to E_y$$
$$v \mapsto s_{\gamma,v}(y),$$

called the **parallel transport along** γ with respect to ∇. This map is invertible, with inverse given by $P_{\gamma^{-1}}$, where $\gamma^{-1}(t) := \gamma(1 - t)$ for each $t \in [0, 1]$.

We can also define P_γ for a (continuous) *piecewise* smooth path γ simply as the composition of the parallel transport maps along its smooth pieces (in the appropriate order). One can show this is well-defined by the uniqueness part of the above cited ODE theorem. Finally, if α is a (piecewise) smooth path starting at $\alpha(0) = y$, then $P_\alpha \circ P_\gamma = P_{\alpha \cdot \gamma}$, where $\alpha \cdot \gamma$ is the concatenation of γ and α:

$$\alpha \cdot \gamma(t) := \begin{cases} \gamma(2t), & \text{if } t \in [0, 1/2], \\ \alpha(2t - 1) & \text{if } t \in [1/2, 1]. \end{cases}$$

The holonomy principle. We can now recall the definition of the holonomy group of ∇.

Definition 1.29 (Holonomy group). Given $x \in M$, the subgroup of $\mathrm{GL}(E_x)$ given by

$$\mathrm{Hol}_x(\nabla) := \{P_\gamma : \gamma \text{ is a piecewise smooth loop based at } x\}$$

is called the **holonomy group** of ∇ at x.

Lemma 1.30. *If $x, y \in M$ are connected by a piecewise smooth path $\gamma \colon [0, 1] \to M$, $\gamma(0) = x$ and $\gamma(1) = y$, then*

$$\mathrm{Hol}_y(\nabla) = P_\gamma \cdot \mathrm{Hol}_x(\nabla) \cdot P_{\gamma^{-1}}.$$

It is easy to see that if $\gamma \colon [0, 1] \to M$ is a *continuous* path connecting $x = \gamma(0)$ and $y = \gamma(1)$, then there exists a *smooth* path $\widetilde{\gamma} \colon [0, 1] \to M$ connecting $x = \widetilde{\gamma}(0)$ and $y = \widetilde{\gamma}(1)$. In fact, we can take $\widetilde{\gamma}$ in the same homotopy class of

γ with fixed end points (see e.g. Kosinski (2007, p. 8, Theorem 2.5)). Thus, the above lemma gives us a precise relation between the holonomy groups of ∇ at any two points lying in the same connected component of M.

If M is connected, we conclude that the holonomy group $\mathrm{Hol}_x(\nabla)$ is independent of the base point x in the following sense. A choice of basis on E_x induces an identification $\mathrm{GL}(E_x) \simeq \mathrm{GL}(r, \mathbb{R})$, and therefore a faithful representation $\mathrm{Hol}_x(\nabla) \hookrightarrow \mathrm{GL}(r, \mathbb{R})$; a different choice of basis will change this identification by conjugation in $\mathrm{GL}(r, \mathbb{R})$. Thus, up to equivalence, there is a well-defined faithful representation of $\mathrm{Hol}_x(\nabla)$ on the typical fibre $\mathbb{R}^r$ of E, called the *holonomy representation*. In this language, the above lemma shows that $\mathrm{Hol}_x(\nabla)$ and $\mathrm{Hol}_y(\nabla)$ have the same holonomy representation. In other words, when regarded as a subgroup of $\mathrm{GL}(r, \mathbb{R})$ defined up to conjugation, the holonomy group is independent of the choice of base point.

Convention 1.31. From now on, we assume M is *connected*, and we write $\mathrm{Hol}(\nabla)$ (omitting the base point), implicitly regarding the holonomy group of ∇ as a subgroup of $GL(r, \mathbb{R})$, defined up to conjugation.

One outcome of the above discussion is that the holonomy group is a *global invariant* of the connection. The next result shows that $\mathrm{Hol}(\nabla)$ 'controls' the existence of $\nabla-$parallel sections ($\nabla t = 0$) on tensors of [14] E Joyce (2007, Proposition 2.5.2, p. 33).

Theorem 1.32 (Holonomy principle). *Let $E \to M$ be a vector bundle over a connected smooth manifold and denote its $(r, s)-$tensor bundle by $\mathcal{T}_s^r(E) := \left(\bigotimes^r E\right) \otimes \left(\bigotimes^s E^*\right)$. Fix a base point $x \in M$, so that $\mathrm{Hol}_x(\nabla)$ acts on E_x, and therefore also on $\mathcal{T}_s^r(E_x)$. Then, any $(r, s)-$tensor $t_x \in \mathcal{T}_s^r(E_x)$ that is invariant under $\mathrm{Hol}_x(\nabla)$ is the value at x of a $\nabla-$parallel $(r, s)-$tensor field $t \in \Gamma\left(\mathcal{T}_s^r(E)\right)$. Conversely, any parallel tensor field $t \in \Gamma\left(\mathcal{T}_s^r(E)\right)$ is fixed in the fibre over x by the action of $\mathrm{Hol}_x(\nabla)$.*

Corollary 1.33. *If $G \subseteq \mathrm{GL}(E_x)$ is the subgroup which fixes $t|_x$ for all parallel tensors t on M, then $\mathrm{Hol}_x(\nabla) \subseteq G$.*

The following result shows that the holonomy group $\mathrm{Hol}(\nabla)$ is a connected Lie group when M is simply-connected Joyce (ibid., Proposition 2.2.4, p. 26).

Proposition 1.34. *Suppose M is simply-connected and ∇ is a connection on a real vector bundle $E \to M$. Then $\mathrm{Hol}(\nabla)$ is a connected Lie subgroup of $\mathrm{GL}(r, \mathbb{R})$.*

---

[14]The statement in Joyce's book is given for $E = TM$ but the proof is clearly valid for any vector bundle $E \to M$

This leads us to consider the notion of *restricted* holonomy groups.

Definition 1.35 (Restricted holonomy group). The **restricted holonomy group** of ∇ at $x \in M$ is the subgroup of $\mathrm{Hol}_x(\nabla)$ given by

$$\mathrm{Hol}_x^0(\nabla) := \{P_\gamma : \gamma \text{ is a null-homotopic piecewise smooth loop based at } x\}.$$

As for the case of the holonomy group, we can regard $\mathrm{Hol}_x^0(\nabla)$ as a subgroup of $\mathrm{GL}(r, \mathbb{R})$, defined up to conjugation, so that we can omit the base point x and write $\mathrm{Hol}^0(\nabla)$. The next proposition gathers some properties of $\mathrm{Hol}^0(\nabla)$ Joyce (ibid., Proposition 2.2.6, p. 27).

Proposition 1.36. $\mathrm{Hol}^0(\nabla)$ *is the connected component of* $\mathrm{Hol}(\nabla)$ *containing the identity and a Lie subgroup of* $\mathrm{GL}(r, \mathbb{R})$. *Moreover, if M is simply-connected then* $\mathrm{Hol}^0(\nabla) = \mathrm{Hol}(\nabla)$.

The Ambrose–Singer theorem. Having in mind the above Proposition:

Definition 1.37 (Holonomy algebra). The **holonomy algebra** $\mathfrak{hol}_x(\nabla)$ of ∇ at $x \in M$ is the Lie algebra of $\mathrm{Hol}_x^0(\nabla)$.

Up to the adjoint action of $\mathrm{GL}(r, \mathbb{R})$, we can also speak of the *holonomy algebra* $\mathfrak{hol}(\nabla)$ as a Lie subalgebra of $\mathfrak{gl}(r, \mathbb{R})$. Actually, the holonomy algebra constrains the curvature F_∇, in the following sense Joyce (ibid., p. 30, Proposition 2.4.1):

Proposition 1.38. *For each $x \in M$, the curvature $F_\nabla|_x$ of ∇ at x lies in* $\Lambda^2 T_x^* M \otimes \mathfrak{hol}_x(\nabla)$.

In fact, a result due to W. Ambrose and I. M. Singer shows that $\mathfrak{hol}(\nabla)$ is *determined* by F_∇ Joyce (ibid., p. 31, Theorem 2.4.3. (a)):

Theorem 1.39 (Ambrose–Singer). *Suppose M is a connected manifold, $E \to M$ is a vector bundle over M, and ∇ is a smooth connection on E. Then, for each $x \in M$, the holonomy algebra $\mathfrak{hol}_x(\nabla)$ is the Lie subalgebra of $\mathrm{End}(E_x)$ spanned, as a vector space, by all elements of $\mathrm{End}(E_x)$ of the form*

$$P_\gamma^{-1}[(F_\nabla)(v, w)]P_\gamma,$$

where $\gamma \colon [0, 1] \to M$ varies on the collection of all piecewise smooth paths starting at $\gamma(0) = x$, and $v, w \in T_{\gamma(1)}M$.

Remark 1.40 (Flat connections). It immediately follows from the Ambrose–Singer theorem that, if ∇ is a *flat* connection, i.e. if $F_\nabla = 0$, then the restricted holonomy group $\mathrm{Hol}_x^0(\nabla)$ is trivial for each $x \in M$. This implies that the parallel transport $P_\gamma \colon E_x \to E_y$ depends only on the homotopy class (with fixed end-points) of the path γ between x and y. In fact, if γ is homotopic to another path $\widetilde{\gamma}$, which without loss of generality we can assume to be piecewise smooth [15], then the concatenation $\widetilde{\gamma}^{-1} \cdot \gamma$ is a null-homotopic (piecewise smooth) loop based at x. Thus, from the triviality of $\mathrm{Hol}_x^0(\nabla)$, we get that $\mathbb{1}_{T_x M} = P_{\widetilde{\gamma}^{-1} \cdot \gamma} = P_{\widetilde{\gamma}}^{-1} \circ P_\gamma$, i.e. $P_\gamma = P_{\widetilde{\gamma}}$, as claimed.

In particular, at every base point $x \in M$, each flat connection on a G–bundle induces a holonomy representation $\pi_1(M, x) \to \mathrm{Aut}(E_x) = G$. Ultimately, this leads to the well-known one-to-one correspondence between gauge-equivalence classes of flat G–connections over M and conjugacy classes of representations $\pi_1(M) \to G$ (cf. Donaldson and Kronheimer (1990, pp. 49-50)). $\Diamond$

1.3 Chern–Weil characteristic classes

Here we give a brief account on the Chern–Weil polynomials representing characteristic classes. This section is based on Milnor (1974, Appendix C).

Fundamental lemma of Chern–Weil theory. Let $E \to M$ be a $\mathbb{K}$–vector bundle of rank r and $\{(U_\alpha, \varphi_\alpha)\}$ an atlas of local trivialisations for E with associated transition functions

$$g_{\alpha\beta} \colon U_\alpha \cap U_\beta \to \mathrm{GL}(r, \mathbb{K}).$$

If ∇ is an arbitrary connection on $E \to M$, its curvature F_∇ is locally described by curvature matrix-valued 2–forms $F_\alpha := [F_\nabla]_\alpha \in \Omega^2(U_\alpha, \mathfrak{gl}(r, \mathbb{K}))$ on M. Compounding the wedge product with matrix multiplication, these objects form a graded multiplicative ring. In particular, we can evaluate a polynomial function $P \colon \mathfrak{gl}(r, \mathbb{K}) \to \mathbb{K}$ on F_α, giving rise to a sum of exterior forms of even degree on U_α.

Now recall from (1.13) that, on overlaps $U_{\alpha\beta} \neq \emptyset$, the F_α and F_β are related by the adjoint action of $\mathrm{GL}(r, \mathbb{K})$: in fact,

$$F_\alpha = g_{\alpha\beta} F_\beta g_{\alpha\beta}^{-1}, \quad \text{on } U_\alpha \cap U_\beta.$$

So if P is a $\mathrm{GL}(r, \mathbb{K})$–**invariant** polynomial, i.e.

$$P(gXg^{-1}) = P(X), \quad \forall g \in \mathrm{GL}(r, \mathbb{K}),$$

---

[15] See the paragraph following Lemma 1.30.

then we can associate a *globally defined* element

$$P(F_\nabla) \in \Omega_{\mathbb{K}}^{\mathrm{even}}(M) := \bigoplus_{k \geqslant 0} \Omega_{\mathbb{K}}^{2k}(M),$$

locally given by:

$$P(F_\nabla)|_{U_\alpha} := P(F_\alpha), \quad \text{for each } \alpha.$$

Of course, in general, $P(F_\nabla)$ will be a sum of exterior forms of various even degrees. But, if we suppose further that P is a **homogeneous** polynomial, i.e. a sum of monomials of a fixed degree m, then $P(F_\nabla) \in \Omega_{\mathbb{K}}^{2m}(M)$. In fact, we could relax the hypothesis of P being invariant to just P being a sum of invariant homogeneous polynomials of increasing degrees, since we know that $Q(F_\nabla) = 0$ whenever $2\deg(Q) > \dim M$.

An important point about $P(F_\nabla)$ is its functorial behaviour, with respect to induced bundles. This means that, if $f : M' \to M$ is a smooth map, then

$$P(f^*\nabla) = f^* P(\nabla).$$

This is a direct consequence of the definition of $f^*\nabla$: if $(U_\alpha, \varphi_\alpha)$ is a local trivialisation of E, then it follows from (1.8) and (1.12) that

$$[F_{f^*\nabla}]_\alpha = f^*[F_\nabla]_\alpha,$$

where $[F_{f^*\nabla}]_\alpha$ denotes the trivialised local form of $F_{f^*\nabla}$.

Crucially, $P(F_\nabla)$ defines a de Rham cohomology class on M, which is independent of the actual connection ∇ on E. This is the content of the next result, which is the core of Chern–Weil theory Milnor (ibid., pp. 296-298):

Proposition 1.41 (Fundamental Chern–Weil Lemma). *Let P be a homogeneous $\mathrm{GL}(r, \mathbb{K})-$invariant polynomial of degree m, and ∇ a connection on a $\mathbb{K}-$vector bundle $E \to M$ of rank r. Then:*

(i) $P(F_\nabla)$ is a closed $2m-$*form, defining a cohomology class $[P(F_\nabla)] \in H_{dR}^{2m}(M, \mathbb{K})$;*

(ii) The class $[P(F_\nabla)]$ is independent of the choice of connection ∇, i.e. if ∇_0 and ∇_1 are connections on E, then the 2m-form $P(F_{\nabla_0}) - P(F_{\nabla_1})$ is exact.

Proof. (i) Let $P'(M) : \mathfrak{gl}(r, \mathbb{K}) \to \mathbb{K}$ be the derivative of P at an element $M \in \mathfrak{gl}(r, \mathbb{K})$. If $X \in \mathfrak{gl}(r, \mathbb{K})$ and $g : \,]-\varepsilon, \varepsilon[\, \to \mathrm{GL}(r, \mathbb{K})$ is given by $g(t) = e^{tX}$, then

$$0 = \frac{\mathrm{d}}{\mathrm{d}t} P(gMg^{-1})|_{t=0} \quad \text{(by the invariance of } P)$$
$$= P'(M)(XM - MX), \tag{1.42}$$

where in the last equation we used the fact that $(gMg^{-1})'(0) = XM - MX$. Now take $M = [F_\nabla]_\alpha \equiv F_\alpha$ and $X = A_\alpha$, where $\nabla|_{U_\alpha} = \mathrm{d} + A_\alpha$ on the local trivialisation φ_α. Then, operating with $\wedge$ in place of the usual multiplication, equation (1.42) reads:

$$P'(F_\alpha) \wedge [A_\alpha, F_\alpha] = 0. \tag{1.43}$$

On the other hand, the chain rule gives

$$\mathrm{d}(P(F_\alpha)) = P'(F_\alpha) \wedge \mathrm{d}F_\alpha. \tag{1.44}$$

Now, the Bianchi identity (1.21) says that $\mathrm{d}F_\alpha + [A_\alpha, F_\alpha] = 0$; therefore,

$$\mathrm{d}(P(F_\alpha)) = -P'(F_\alpha) \wedge [A_\alpha, F_\alpha] = 0,$$

as claimed.

(ii) Suppose ∇_0 and ∇_1 are different connections on E. Write $p \colon M \times \mathbb{R} \to M$ for the canonical projection, and consider the induced connections $\nabla'_l := p^*\nabla_l, l = 0, 1$, on $p^*E \to M$ and their convex combination

$$\nabla := t\nabla'_1 + (1 - t)\nabla'_0,$$

where $t \colon M \times \mathbb{R} \to \mathbb{R}$ is the natural projection. If $i_l \colon M \to M \times \mathbb{R}$ denotes the function $x \mapsto (x, l), l = 0, 1$, then we can identify $i_l^* \nabla$ with $\nabla_l, l = 0, 1$, as connections on E. Being a smooth map, it follows that

$$i_l^*(P(F_\nabla)) = P(F_{\nabla_l}).$$

Now, the maps i_0 and i_1 are clearly homotopic, so they induce the same map in cohomology. In particular,

$$P(F_{\nabla_0}) = i_0^*(P(F_\nabla)) = i_1^*(P(F_\nabla)) = P(F(\nabla_1))⊐$$

In summary, each invariant homogeneous polynomial P on $\mathfrak{gl}(r, \mathbb{K})$ determines a *characteristic cohomology class* $c_P(E) = [P(F_\nabla)]$ in $H_{dR}^*(M, \mathbb{K})$, depending only on the isomorphism class of the vector bundle E, and such that, if $f \colon M' \to M$ is smooth, then

$$c_P(f^*E) = f^*c_P(E).$$

Here the left-hand side represents the cohomology class of the pull-back bundle f^*E and the right-hand side is the image of the cohomology class associated to E under the pull-back map induced by f in cohomology.

Chern classes in de Rham cohomology. Let $\mathbb{K} = \mathbb{C}$. For each $X \in \mathfrak{gl}(r, \mathbb{C})$ and $1 \leqslant k \leqslant r$, write $\sigma_k(X)$ for the k–th elementary symmetric polynomial function on the eigenvalues of X, so that

$$\det(\mathbb{1} + tX) = 1 + t\sigma_1(X) + \ldots + t^r \sigma_r(X).$$

More explicitly, if $\lambda_1, \ldots, \lambda_r \in \mathbb{C}$ are the eigenvalues of $X \in \mathfrak{gl}(r, \mathbb{C})$, then

$$\sigma_k(X) = \sum_{1 \leqslant i_1 < \ldots < i_k \leqslant r} \lambda_{i_1} \ldots \lambda_{i_k},$$

for each $1 \leqslant k \leqslant r$. In particular, $\sigma_1(X) = \operatorname{tr} X$ and $\sigma_r(X) = \det X$.

Every symmetric polynomial $P : \mathfrak{gl}(r, \mathbb{C}) \to \mathbb{C}$ has a unique representation as a polynomial in these elementary functions $\sigma_1, \ldots, \sigma_r$. From this, one has Milnor (1974, p. 299, Lemma 6):

Proposition 1.45. *The ring of* $\mathrm{GL}(r, \mathbb{C})$*–invariant polynomials is* $\mathbb{C}[\sigma_1, \ldots, \sigma_r]$*, i.e. every invariant polynomial on* $\mathfrak{gl}(r, \mathbb{C})$ *can be expressed as a polynomial function of* $\sigma_1, \ldots, \sigma_r$.

Definition 1.46 (Chern classes and Chern character in de Rham cohomology). Let $E \to M$ be a complex vector bundle of rank r and let ∇ be a smooth connection on E. For $1 \leqslant k \leqslant r$,

1. the k-**th Chern class** of E is the element:

$$c_k(E) := \left(\frac{-1}{2\pi \mathbf{i}} \right)^k [\sigma_k (F_\nabla)] \in H^{2k}_{dR}(M, \mathbb{C}).$$

2. the k-**th Chern character** of E is the element:

$$\mathrm{ch}_k(E) := \frac{(-1)^k}{(2\pi \mathbf{i})^k k!} [\operatorname{tr}(F_\nabla \wedge \ldots \wedge F_\nabla)] \in H^{2k}_{dR}(M, \mathbb{C}).$$

For instance, the first two Chern classes are represented as follows:

$$c_1(E) = \frac{\mathbf{i}}{2\pi}[\operatorname{tr}(F_\nabla)] \tag{1.47}$$

and

$$c_2(E) = \frac{-1}{8\pi^2}[\operatorname{tr}(F_\nabla) \wedge \operatorname{tr}(F_\nabla) - \operatorname{tr}(F_\nabla \wedge F_\nabla)]. \tag{1.48}$$

1.4 Yang–Mills equation on Riemannian manifolds

In this section we review the variational formulation of the (weak/strong) Yang–Mills equation on a Riemannian $n-$manifold, by means of the Yang–Mills energy functional, and we point out some of its basic symmetries. The references for this section are Wehrheim (2004, pp. 141-142, 172–173) and Uhlenbeck (1982b, §1).

Yang–Mills functional. Let (M, g) be an oriented Riemannian $n-$manifold and let $E \to M$ be a $G-$bundle. Denote by $\langle \cdot, \cdot \rangle$ the natural tensor product metric on $\Lambda^2 T^* M \otimes \mathfrak{g}_E$ induced by g and the Ad_G-invariant inner product (1.1) on $\mathfrak{g}$. Then, for each $\xi, \zeta \in \Omega^2(M, \mathfrak{g}_E)$, we have

$$\langle \xi, \zeta \rangle \mathrm{d}V_g = \langle \xi \wedge *\zeta \rangle_{\mathfrak{g}} = -\mathrm{tr}(\xi \wedge *\zeta),$$

where $\langle \xi \wedge *\zeta \rangle_{\mathfrak{g}}$ represents the contraction of $\xi \wedge *\zeta$ by the induced invariant metric on $\mathfrak{g}_E$.

If $|\cdot|$ denotes the induced pointwise norm on sections of $\Lambda^2 T^* M \otimes \mathfrak{g}_E$, then for each $\nabla \in \mathcal{A}(E)$ we get a function $|F_\nabla| \colon M \to \mathbb{R}$. By the Ad_G-invariance of $\langle \cdot, \cdot \rangle_{\mathfrak{g}}$ and (1.23), it follows that

$$|F_{g*\nabla}| = |F_\nabla|, \quad \text{for each } g \in \mathcal{G}(E) \text{ and } \nabla \in \mathcal{A}(E). \tag{1.49}$$

In other words, the function $\nabla \mapsto |F_\nabla|$ is invariant under the action of $\mathcal{G}(E)$ (*gauge invariant*).

Definition 1.50. The **Yang–Mills functional**

$$\mathcal{YM} \colon \mathcal{A}(E) \to [0, \infty]$$

associates to each connection $\nabla \in \mathcal{A}(E)$ its L^2-**energy**:

$$\mathcal{YM}(\nabla) := \|F_\nabla\|_{L^2}^2 = \int_M |F_\nabla|^2 \mathrm{d}V_g = -\int_M \mathrm{tr}(F_\nabla \wedge *F_\nabla).$$

If the base manifold M is not compact, the L^2-energy of a smooth connection ∇ might be infinite, and natural Sobolev spaces of connections might be hard to define. On the other hand, if M is compact, then $\mathcal{YM}$ is clearly finite on the whole space of smooth connections $\mathcal{A}(E)$. Moreover, one can prove that $\mathcal{YM}$ extends to $\mathcal{A}^{1,p}(E)$, for each $2 \leqslant p < \infty$ such that [16] $p \geqslant \frac{4n}{4+n}$.

---

[16] The latter condition comes from the Sobolev embedding $W^{1,p} \hookrightarrow L^4$ (Theorem B.13) which ensures that $[A, A]$ and hence $F_\nabla = F_{\nabla_0} + \mathrm{d}_{\nabla_0} A + [A, A]$ lie in L^2, whenever $\nabla = \nabla_0 + A \in \mathcal{A}^{1,p}(E)$.

It follows directly from (1.49) that $\mathcal{YM}$ is a gauge-invariant functional on $\mathcal{A}(E)$, i.e.

$$\mathcal{YM}(g^*\nabla) = \mathcal{YM}(\nabla), \quad \text{for each } g \in \mathcal{G}(E) \text{ and } \nabla \in \mathcal{A}(E). \tag{1.51}$$

Furthermore, $\mathcal{YM}$ is *conformally invariant* if, and only if, $n = 4$. Indeed, if we scale g by some positive smooth function f on M, then the pointwise inner product on 2–forms scales by f^{-2}, while the Riemannian volume n–form scales by $f^{n/2}$. Thus, an integral $\int_M |F_\nabla|^2 dV_g$ transforms to $\int_M f^{\frac{n}{2}-2}|F_\nabla|^2 dV_g$, which stays invariant precisely when $n = 4$.

Yang–Mills equation. Let us derive the first variational formula of $\mathcal{YM}$, with respect to compactly supported variations:

Proposition 1.52. *Let $\nabla \in \mathcal{A}(E)$ with $\mathcal{YM}(\nabla) < \infty$. If $\{\nabla_t\}_{t\in]-\varepsilon,\varepsilon[}$ is a compactly supported smooth variation of ∇, then*

$$\frac{d}{dt}\mathcal{YM}(\nabla_t)|_{t=0} = 2\langle d^*_\nabla F_\nabla, B\rangle_{L^2},$$

where [17]

$$B := \frac{d}{dt}\nabla_t\Big|_{t=0} \in \Gamma_0(T^*M \otimes \mathfrak{g}_E).$$

In particular, ∇ is a critical point of $\mathcal{YM}$ with respect to compactly supported smooth variations if, and only if, ∇ satisfies the (strong) **Yang–Mills equation** [18].

$$d^*_\nabla F_\nabla = 0. \tag{1.53}$$

Proof. Recall that a smooth variation of ∇ is just a smooth path $t \mapsto \nabla_t$ on $\mathcal{A}(E)$ starting at $\nabla_0 = \nabla$. We say that a smooth variation $\{\nabla_t\}$ is compactly supported provided there exists a precompact open subset $U \Subset M$ such that, writing $\nabla_t = \nabla + A_t$, where $A_t \in \Omega^1(M, \mathfrak{g}_E)$, then each A_t has compact support contained in $\overline{U}$. The statement that ∇ is a critical point of $\mathcal{YM}$ with respect to compactly supported variations means simply that $\frac{d}{dt}\mathcal{YM}(\nabla_t)|_{t=0} = 0$ for all such variations.

---

[17]Henceforth, $\Gamma_0(\cdot) \subset \Gamma(\cdot)$ denotes the subset of compactly supported sections.

[18]If M is a compact manifold with (possibly empty) boundary ∂M, the Yang–Mills equation becomes the system:

$$\begin{cases} d^*_\nabla F_\nabla = 0 \\ *F_\nabla|_{\partial M} = 0. \end{cases}$$

By the affine space structure of $\mathcal{A}(E)$, we may restrict ourselves to variations of the form $\nabla_t = \nabla + tB$, where $B \in \Gamma_0(T^*M \otimes \mathfrak{g}_E)$. In this case, locally, we have:

$$[F_{\nabla_t}]_\alpha = \mathrm{d}(A_\alpha + tB) + (A_\alpha + tB) \wedge (A_\alpha + tB)$$
$$= F_\alpha + t(\mathrm{d}B + B \wedge A_\alpha + A_\alpha \wedge B) + t^2(B \wedge B)$$
$$= [F_\nabla + t(\mathrm{d}_\nabla B) + t^2(B \wedge B)]_\alpha.$$

Hence, globally,

$$F_{\nabla_t} = F_\nabla + t(\mathrm{d}_\nabla B) + t^2(B \wedge B).$$

In particular,

$$\frac{\mathrm{d}}{\mathrm{d}t}\mathcal{YM}(\nabla_t)|_{t=0} = \frac{\mathrm{d}}{\mathrm{d}t}\langle F_{\nabla_t}, F_{\nabla_t}\rangle_{L^2}|_{t=0}$$
$$= 2\left\langle \frac{\mathrm{d}}{\mathrm{d}t}F_{\nabla_t}|_{t=0}, F_\nabla \right\rangle_{L^2}$$
$$= 2\langle \mathrm{d}_\nabla B, F_\nabla\rangle_{L^2}$$
$$= 2\langle B, \mathrm{d}_\nabla^* F_\nabla\rangle_{L^2},$$

where in the second equality we use [19] $\mathcal{YM}(\nabla) < \infty$, and the last equality follows from the definition of d_∇^* as a formal L^2−adjoint of d_∇ (provided the compact support of B does not intersect a nonempty ∂M). $\qquad\square$

Remark 1.54. Since $\mathrm{d}_\nabla^* = \pm * \mathrm{d}_\nabla*$, the Yang–Mills equation can be rewritten as

$$\mathrm{d}_\nabla * F_\nabla = 0.$$

This condition does *not* depend on the choice of orientation on M; indeed, a change in orientation only causes the $*$−operator to change sign, clearly not affecting the equation. $\qquad\Diamond$

Definition 1.55 (Yang–Mills connections). A smooth connection $\nabla \in \mathcal{A}(E)$ satisfying the Yang–Mills equation (1.53) is called a **Yang–Mills connection**; its curvature tensor is called a **Yang–Mills field**.

Another important class of connections consists of *weak* Yang–Mills connections, which are critical points of the Yang–Mills action on appropriate Sobolev spaces:

---

[19]Indeed, we need this to ensure that F_{∇_t} is L^2−integrable so we can apply the standard theorem of differentiation under the integral sign, e.g. as in Folland (2013, Theorem 2.27).

Definition 1.56 (Weak Yang–Mills connections). Suppose M^n is compact. Let $1 \leqslant p < \infty$ be such that $p > \frac{n}{2}$, and if $n = 2$ assume moreover $p \geqslant \frac{4}{3}$. A connection $\nabla \in \mathcal{A}^{1,p}(E)$ is called a **weak Yang–Mills connection** when it satisfies the **weak Yang–Mills equation**:

$$\int_M \langle F_\nabla, \mathrm{d}_\nabla B \rangle \mathrm{d}V_g = 0, \quad \forall B \in \Gamma_0(T^*M \otimes \mathfrak{g}_E). \tag{1.57}$$

Remark 1.58. The Yang–Mills functional need not be finite, nor even defined, on weak Yang–Mills connections. The assumptions on p, stemming from Sobolev embedding (Theorem B.13), ensure both that the weak Yang–Mills equation makes sense for those connections and that the equation is preserved by the action of $\mathcal{G}^{2,p}(E)$ (the latter requires the strict inequality $p > \frac{n}{2}$). Moreover, it is not a priori clear whether weak Yang–Mills connections satisfy the strong Yang–Mills equation, although this is true for sufficiently regular connections (see e.g. Wehrheim (2004, Lemma 9.3, p. 142)). $\Diamond$

We note that both the weak and strong Yang–Mills equations are invariant under gauge transformations. For the weak equation this is more subtle, and we refer the reader to Wehrheim (ibid., Lemma 9.2, p. 142). For the strong equation, in the light of Proposition 1.52, one can deduce this fact from the invariance (1.51) of the Yang–Mills functional on $\mathcal{A}(E)$. Alternatively, one can check directly that, for each $\nabla \in \mathcal{A}(E)$ and $g \in \mathcal{G}(E)$,

$$\mathrm{d}_{g*\nabla} * F_{g*\nabla} = g^{-1}(\mathrm{d}_\nabla * F_\nabla)g.$$

In particular, the solutions of the Yang–Mills equation, seen as either Yang–Mills connections or fields, are an invariant set under gauge transformations. This *gauge freedom* turns out to be the main difficulty in treating the regularity theory of these equations.

Gauge fixing. Our reference for this section is the landmark article Uhlenbeck (1982b, §1). Let us take a closer look at the Yang–Mills equation (1.53), in a local gauge $(U_\alpha, \varphi_\alpha)$ of the G–bundle E. Suppose we have coordinates $(x^1, \ldots, x^n)$ on U_α; write $\nabla_\alpha = \mathrm{d} + A_\alpha$, $F_\alpha = \mathrm{d}A_\alpha + A_\alpha \wedge A_\alpha$ and recall the local expressions (1.14) and (1.15). For simplicity, assume also that $(g_{ij}) = (\delta_{ij})$, i.e. g is *flat* on U_α. Then,

$$\mathrm{d}_\nabla^* F_\nabla = \mathrm{d}^* F_\alpha - *[A_\alpha, *F_\alpha]$$

and

$$\mathrm{d}^* F_\alpha = -\sum_{i,j} \frac{\partial F_{\alpha,ij}}{\partial x^i} \otimes \mathrm{d}x^j.$$

Hence

$$\mathrm{d}_\nabla^* F_\nabla = -\sum_{i,j} \left(\frac{\partial F_{\alpha,ij}}{\partial x^i} + [A_{\alpha,i}, F_{\alpha,ij}] \right) \otimes \mathrm{d}x^j.$$

Therefore, in the gauge $(U_\alpha, \varphi_\alpha)$, equation (1.53) reads

$$\sum_i \frac{\partial F_{\alpha,ij}}{\partial x^i} + [A_{\alpha,i}, F_{\alpha,ij}] = 0, \quad \forall j = 1, \ldots, n.$$

Of course, for general g, a similar (more complicated) equation holds. Now, if the gauge group G is an Abelian Lie group (and therefore all brackets on $\mathfrak{g}$ are zero; in particular, $F_\alpha = \mathrm{d}A_\alpha$), the Bianchi identity and the Yang–Mills equation for ∇_α reduce to $\mathrm{d}F_\alpha = \mathrm{d}^2 A_\alpha = 0$ and $\mathrm{d}^* F_\alpha = 0$, respectively. This pair of equations then forms an **elliptic system** for F_α. This is the basic linear model for the regularity theory.

In the non-Abelian case, a non-smooth gauge transformation g can turn a smooth field F_α into a discontinuous field $g F_\alpha g^{-1}$. Thus, the choice of a 'good' gauge is much more important to the non-linear theory. The linearised Yang–Mills equations for A_α are $\mathrm{d}^* \mathrm{d} A_\alpha = 0$. By the last paragraph, this is exactly the Yang–Mills equation if G is Abelian. This single system for A_α is *not* elliptic and, just as in Hodge theory for exact forms on manifolds, one usually adds a second equation such as $\mathrm{d}^* A_\alpha = 0$ to remedy the situation. In the Abelian case, this involves solving the linear equation $\mathrm{d}^*(A_\alpha + \mathrm{d}u) = \mathrm{d}^* \widetilde{A}_\alpha = 0$ for $u \colon U_\alpha \to \mathfrak{g}$. Here $\widetilde{A}_\alpha := g^* A_\alpha = g^{-1} A_\alpha g + g^{-1} \mathrm{d}g$, where $g = e^u \in C^\infty(U_\alpha, G) \simeq \mathcal{G}(E|_{U_\alpha})$.

The equation $\mathrm{d}^* A_\alpha = 0$ can also be added to the non-linear theory as a method of *choosing a good gauge*. In general, to find such a gauge we need to solve the non-linear elliptic equation: $\mathrm{d}^*(g^{-1} A_\alpha g + g^{-1} \mathrm{d}g) = 0$ for $g \in C^\infty(U_\alpha, G)$. Such a solution is often called a *Coulomb gauge*. In the seminal works Uhlenbeck (1982b, Theorems 2.7 and 2.8) and Uhlenbeck (1982a, Theorem 1.3), K. Uhlenbeck solves the general problem of constructing Coulomb gauges over model domains of interest under, respectively, L^∞ and $L^{n/2}$–boundedness hypothesis on the curvature norm. In Section 3.1, we will see more about the latter (cf. Theorem 3.3).

Finally, notice that one can also study the Yang–Mills equation on Lorentzian manifolds, its original formulation from Physics, as a generalisation of Maxwell's equations on Minkowski space-time $\mathbb{R}^{3+1}$. The resulting equation is weakly hyperbolic, and it turns out to be very hard to analyse.

1.5 Instantons in four dimensions

Let us illustrate the previous concepts in action, by reviewing the basic aspects of the classical 4–dimensional theory. The main references are, of course, Donaldson and Kronheimer (1990, §2.1.3-2.1.5) and Freed and Uhlenbeck (1984, pp. 36-37), see also Scorpan (2005, Chapter 9, pp. 351-354).

Let (M^4, g) be an oriented Riemannian 4–manifold. A special feature of this setting is that the Hodge star operator on 2–forms,

$$*\colon \Lambda^2 T^* M \to \Lambda^2 T^* M,$$

is an involutive [20] self-adjoint [21] automorphism. Hence, $*|_{\Lambda^2 T^* M}$ has eigenvalues ± 1 and it splits $\Lambda^2 T^* M$ orthogonally into the corresponding eigenbundles $\Lambda_\pm^2 T^* M$:

$$\Lambda^2 T^* M = \Lambda_+^2 T^* M \oplus \Lambda_-^2 T^* M, \tag{1.59}$$

where $\Lambda_\pm^2 T^* M := \{\omega \in \Lambda^2 T^* M : *\omega = \pm\omega\}$. Fibrewise, this phenomenon corresponds to the exceptional Lie algebra isomorphism

$$\mathfrak{so}(4) \simeq \mathfrak{so}(3) \oplus \mathfrak{so}(3),$$

which, at the level of Lie groups, reads

$$\mathrm{Spin}(4) = \mathrm{SU}(2) \times \mathrm{SU}(2).$$

Indeed, as SO(4)–modules, one has $\Lambda^2(\mathbb{R}^4)^* \simeq \mathfrak{so}(4)$, and this isomorphism maps the $*$–eigenspaces Λ_+^2 and Λ_-^2 onto the two 3–dimensional commuting ideals in $\mathfrak{so}(4)$, isomorphic to $\mathfrak{so}(3)$. We note that the SO(4)–modules Λ_+^2 and Λ_-^2 are both irreducible and 3–dimensional, but not SO(4)–isomorphic Besse (2008, §1.123-1.125, p. 50).

Defining $\Omega_\pm^2(M) := \Gamma(\Lambda_\pm^2 T^* M)$, we get an L^2–orthogonal decomposition of $\Omega^2(M)$ as

$$\Omega^2(M) = \Omega_+^2(M) \oplus \Omega_-^2(M).$$

Accordingly, every $\omega \in \Omega^2(M)$ can be written as $\omega = \omega^+ \oplus \omega^-$, with

$$\omega^\pm := \frac{\omega \pm *\omega}{2} \in \Omega_\pm^2(M).$$

---

[20] In general, $*^2 = (-1)^{k(4-k)}\mathbb{1}$ when acting on $\Lambda^k T^* M$ (see e.g. Petersen (2006, Lemma 26, p. 203)).

[21] with respect to the natural metric $(\cdot, \cdot)_g$ on $\Lambda^2 T^* M$ induced by g; recall that $*$ is defined as the unique bundle isomorphism $\Lambda^k T^* M \simeq \Lambda^{n-k} T^* M$ such that $\alpha \wedge *\beta = (\alpha, \beta)_g \, dV_g$ for all $\alpha, \beta \in \Lambda^k T^* M$.

Definition 1.60. A 2–form $\omega \in \Omega^2(M)$ is called **anti-selfdual** (resp. **selfdual**) if $\omega^+ = 0$ (resp. $\omega^- = 0$). We adopt the obvious abbreviations **(A)SD**.

Remark 1.61. A change of orientation on M changes the Hodge star operator $*$ by a sign and thus reverses the roles of $\Lambda^2_+ T^*M$ and $\Lambda^2_- T^*M$. Moreover, as the action of $*$ on 2–forms in dimension 4 is conformally invariant, the (A)SD condition is also conformally invariant. $\Diamond$

Given a G–bundle E over M, the L^2–orthogonal splitting (1.59) immediately extends to $\mathfrak{g}_E$–valued 2–forms:

$$\Omega^2(M, \mathfrak{g}_E) = \Omega^2_+(M, \mathfrak{g}_E) \oplus \Omega^2_-(M, \mathfrak{g}_E),$$

where $\Omega^2_\pm(M, \mathfrak{g}_E) := \Gamma(\Lambda^2_\pm T^*M \otimes \mathfrak{g}_E)$. For $\nabla \in \mathcal{A}(E)$, we write

$$F_\nabla = F_\nabla^+ \oplus F_\nabla^- \in \Omega^2_+(M, \mathfrak{g}_E) \oplus \Omega^2_-(M, \mathfrak{g}_E).$$

This gives rise to a very important class of solutions for the Yang–Mills equation in four dimensions.

Definition 1.62. Let (M, g) be an oriented Riemannian 4–manifold and let $E \to M$ be a G–bundle with compact semi-simple structure group [22]. A smooth connection $\nabla \in \mathcal{A}(E)$ is called an **ASD** (resp. **SD**) **instanton** when $F_\nabla^+ = 0$ (resp. $F_\nabla^- = 0$).

A few observations are in order.

- The (A)SD equation $*F_\nabla = \pm F_\nabla$ is both *gauge invariant* and *conformally invariant*. For the gauge invariance, note that if $\nabla \in \mathcal{A}(E)$ is an (A)SD instanton and $g \in \mathcal{G}(E)$, then

$$*F_{g*\nabla} = *(g^{-1} F_\nabla g) = g^{-1}(*F_\nabla)g = \pm F_{g*\nabla}.$$

  As to conformal invariance, it follows from the conformal invariance of the Hodge star $*$ on 2–forms in four dimensions.

- *Every (A)SD instanton is a Yang–Mills connection.* Indeed, by the Bianchi identity (1.21):

$$*F_\nabla = \pm F_\nabla \quad \Rightarrow \quad \mathrm{d}_\nabla(*F_\nabla) = \pm \mathrm{d}_\nabla F_\nabla = 0.$$

  Notice also that the (A)SD equation $F_\nabla^\pm = 0$ is a (nonlinear, unless G is Abelian) *first-order* p.d.e. on the connection, while the Yang–Mills

---

[22]In Chapter 2 we will extend this definition allowing G to be any compact Lie group; see Definition 2.80 (ii) and the subsequent discussion.

equation $d_\nabla^* F_\nabla = 0$ is a (nonlinear, unless G is Abelian) *second-order* p.d.e. on the connection. One moral is that (A)SD instantons provide a fertile source of examples of Yang–Mills connections. Nonetheless, one can construct examples of Yang–Mills connections which are neither SD nor ASD. For instance, L. M. Sibner, R. J. Sibner, and Uhlenbeck (1989) were the first to give such examples, for $M = S^4$ and $G = \mathrm{SU}(2)$; two years later, Sadun and Segert (1991) showed that non-selfdual Yang–Mills connections exist on all $\mathrm{SU}(2)-$bundles over S^4 with second Chern number [23] not equal to ± 1. See also Wang (1991), for examples on $M = S^3 \times S^1$ (and $S^2 \times S^2$) with $G = \mathrm{SU}(2)$.

Topological energy bounds from Chern–Weil theory. Suppose M is a closed oriented Riemannian 4–manifold and let $E \to M$ be an $\mathrm{SU}(r)-$bundle. We will show that $\mathcal{YM}\colon \mathcal{A}(E) \to \mathbb{R}$ is bounded below by a number depending only on the topology of E. Furthermore, the sign of such lower bound obstructs the existence of either SD or ASD instantons on E, which are shown to be the *absolute minima* of $\mathcal{YM}$.

Given $\nabla \in \mathcal{A}(E)$, by the basic Chern–Weil theory from Section 1.3, we know that the topological characteristic class $c_2(E)$ is represented by

$$
\begin{aligned}
c_2(E) &= -\frac{1}{8\pi^2}[\mathrm{tr}(F_\nabla) \wedge \mathrm{tr}(F_\nabla) - \mathrm{tr}(F_\nabla \wedge F_\nabla)] \\
&= \frac{1}{8\pi^2}[\mathrm{tr}(F_\nabla \wedge F_\nabla)]. \quad \text{(since } F_\nabla \in \Omega^2(M, \mathfrak{su}_E) \text{ is trace-free)}
\end{aligned}
$$

We define the **topological charge** $\kappa(E)$ of E by pairing $c_2(E)$ with the fundamental class $[M]$:

$$
\kappa(E) := \langle c_2(E), [M] \rangle = \frac{1}{8\pi^2} \int_M \mathrm{tr}(F_\nabla \wedge F_\nabla).
$$

From the L^2-orthogonal decomposition

$$
F_\nabla = F_\nabla^+ \oplus F_\nabla^- \in \Omega_+^2(M, \mathfrak{g}_E) \oplus \Omega_-^2(M, \mathfrak{g}_E),
$$

it follows that

$$
\begin{aligned}
8\pi^2 \kappa(E) &= -\langle F_\nabla, *F_\nabla \rangle_{L^2} \\
&= -\langle F_\nabla^+ + F_\nabla^-, F_\nabla^+ - F_\nabla^- \rangle_{L^2} \\
&= -\|F_\nabla^+\|_{L^2}^2 + \|F_\nabla^-\|_{L^2}^2.
\end{aligned}
$$

---

[23]The second Chern number of a complex vector bundle $E \to M$ over an oriented compact 4–manifold is the integer $C_2(E)$ given by the natural pairing $\langle c_2(E), [M] \rangle$.

                                              1. Geometry and gauge theory

On the other hand,

$$\mathcal{YM}(\nabla) = \|F_\nabla\|_{L^2}^2 = \|F_\nabla^+\|_{L^2}^2 + \|F_\nabla^-\|_{L^2}^2.$$

Thus we get two identities:

$$\mathcal{YM}(\nabla) = 2\|F_\nabla^\pm\|_{L^2}^2 \pm 8\pi^2 \kappa(E).$$

In particular, $\mathcal{YM}(\nabla) \geqslant 8\pi^2 |\kappa(E)|$, and we distinguish the following cases:

- if $\kappa(E) = 0$ then the absolute minima for $\mathcal{YM}$ are precisely the (A)SD *flat* connections;

- if $\kappa(E) > 0$ then E does not admit SD instantons and
  ∇ is an **absolute minima** for $\mathcal{YM}$ $\iff$ $\mathcal{YM}(\nabla) = 8\pi^2 \kappa(E)$ $\iff$
  ∇ is an **ASD** instanton;

- if $\kappa(E) < 0$ then E does not admit ASD instantons and
  ∇ is an absolute minima for $\mathcal{YM}$ $\iff$ $\mathcal{YM}(\nabla) = -8\pi^2 \kappa(E)$ $\iff$
  ∇ is an SD instanton.

ASD instantons on $\mathbb{R}^4$.   Here we present some concrete constructions, drawn from Jardim (2005, §2.3), Martino (2011, §3.2) and G. L. Naber (2011, §6.3).

Consider the Euclidean space $M = \mathbb{R}^4$, with its standard oriented Riemannian manifold structure, and let $E \to M$ be a (necessarily) trivial G−bundle with compact semi-simple structure group. In Euclidean coordinates $x^1, \ldots, x^n$, any connection $\nabla \in \mathcal{A}(E)$ can be written globally as $\nabla = \mathrm{d} + A$, for some

$$A = \sum_{i=1}^4 A_i \otimes \mathrm{d}x^i, \quad A_i : \mathbb{R}^4 \to \mathfrak{g}.$$

Furthermore, from (1.16),

$$F_\nabla = \frac{1}{2} \sum F_{ij} \otimes \mathrm{d}x^i \wedge \mathrm{d}x^j,$$

with

$$F_{ij} = \partial_i A_j - \partial_j A_i + [A_i, A_j].$$

In this context, we have explicitly:

$$\nabla \text{ is an ASD instanton} \quad \iff \quad \begin{cases} F_{12} + F_{34} &= 0 \\ F_{13} + F_{42} &= 0 \\ F_{14} + F_{23} &= 0 \end{cases}. \tag{1.63}$$

The first non-trivial explicit examples of ASD instantons on $\mathbb{R}^4$, with finite L^2–energy and gauge group $G = \mathrm{SU}(2) \simeq \mathrm{Sp}(1)$, were given in the classical paper Belavin et al. (1975). The simplest solution, called the *basic instanton*, has the potential

$$A(x) := \frac{1}{|x|^2 + 1} \mathrm{Im}(q \, \mathrm{d}\overline{q}),$$

where q is the quaternion $x^1 + x^2\mathbf{i} + x^3\mathbf{j} + x^4\mathbf{k}$, while $\mathrm{Im}(q\,\mathrm{d}\overline{q})$ denotes the imaginary part of the product quaternion $q\,\mathrm{d}\overline{q}$. Here we are regarding $\mathbf{i}, \mathbf{j}, \mathbf{k}$ as a basis of the Lie algebra $\mathfrak{su}(2) \simeq \mathrm{Im}(\mathbb{H})$. Computing the curvature of $\nabla = \mathrm{d} + A$, one gets

$$F_\nabla(x) := \frac{1}{\left(|x|^2 + 1\right)^2} \mathrm{d}q \wedge \mathrm{d}\overline{q}.$$

Note that the action density function

$$|F_\nabla|^2(x) = \frac{48}{\left(|x|^2 + 1\right)^4}$$

has a bell-shaped profile, centred at the origin and decaying like r^{-8}. Furthermore, one can show that ∇ has *topological charge* 1, i.e.

$$\begin{aligned}
\kappa(\nabla) &:= \frac{1}{8\pi^2} \int_M \mathrm{tr}(F_\nabla \wedge F_\nabla) \\
&= \frac{1}{8\pi^2} \int_{\mathbb{R}^4} \frac{48}{\left(|x|^2 + 1\right)^4} \mathrm{d}x \\
&= \frac{1}{8\pi^2} \mathrm{Vol}(S^3) \int_0^\infty \frac{48r^3}{\left(r^2 + 1\right)^4} \mathrm{d}r = 1.
\end{aligned}$$

More generally, given $x_0 \in \mathbb{R}^4$ and $\lambda \in \mathbb{R}_+$, denoting by $t_{x_0,\lambda} : \mathbb{R}^4 \to \mathbb{R}^4$ the isometry

$$t_{x_0,\lambda}(x) := \lambda^{-1}(x - x_0), \quad \forall x \in \mathbb{R}^4,$$

then the pull-back connection $\nabla_{x_0,\lambda} := t^*_{x_0,\lambda} \nabla$ is still an ASD instanton; more explicitly, letting x_0 correspond to the quaternion q_0, we can write

$$A_{x_0,\lambda}(x) = \frac{1}{|x - x_0|^2 + \lambda^2} \mathrm{Im}\left((q - q_0)\mathrm{d}\overline{q}\right)$$

and

$$F_{\nabla_{x_0,\lambda}}(x) = \frac{\lambda^2}{\left(|x - x_0|^2 + \lambda^2\right)^2} \mathrm{d}q \wedge \mathrm{d}\overline{q}.$$

The action density function

$$|F_{\nabla_{x_0,\lambda}}|^2(x) = \frac{48\lambda^4}{\left(|x - x_0|^2 + \lambda^2\right)^4}$$

has a bell-shaped profile centred at x_0, and one still has

$$\kappa(\nabla_{x_0,\lambda}) = \frac{1}{8\pi^2}\mathcal{YM}(\nabla_{x_0,\lambda})$$

$$= \frac{1}{8\pi^2}\mathrm{Vol}(S^3)\int_0^\infty \frac{48\lambda^4 r^3}{\left(r^2 + \lambda^2\right)^4}\,dr = 1. \quad \text{(indep. of } x_0 \text{ and } \lambda)$$

On the other hand, for fixed x_0,

$$\sup_{x \in \mathbb{R}^4} |F_{\nabla_{x_0,\lambda}}|^2(x) = |F_{\nabla_{x_0,\lambda}}|^2(x_0) = \lambda^{-4}48 \xrightarrow{\lambda \downarrow 0} \infty.$$

Thus, as $\lambda \downarrow 0$, the action density function $|F_{\nabla_{x_0,\lambda}}|^2$ concentrates more and more at x_0. We shall refer to x_0 as the *centre* and λ as the *scale* of the potential $A_{x_0,\lambda}$.

Instantons of topological charge k, also called *pseudoparticles*, can be obtained by "superimposing" k basic instantons, via the so-called *'t Hooft Ansatz*. Given $y_i \in \mathbb{R}^4$ and $\lambda_i \in \mathbb{R}_+$, $i = 1, \ldots, k$, consider the positive harmonic function $\rho \colon \mathbb{R}^4 \to \mathbb{R}$ given by

$$\rho(x) := 1 + \sum_{i=1}^k \frac{\lambda_i^2}{|x - y_i|^2}.$$

Then the potential $A = A_\mu \otimes dx^\mu$, with

$$A_\mu := \mathbf{i} \sum_\nu \sigma_{\mu\nu} \frac{\partial}{\partial x^\nu} \ln(\rho),$$

defines an ASD instanton; here, $\sigma_{\mu\nu}$, $\mu, \nu = 1, 2, 3, 4$, are the skew-symmetric matrices

$$\sigma_{jl} := \frac{1}{4\mathbf{i}}[\sigma_j, \sigma_l], \quad \sigma_{j4} := \frac{1}{2}\sigma_j, \quad j, l = 1, 2, 3,$$

where $\sigma_1, \sigma_2, \sigma_3$ are the Pauli matrices. This is interpreted as a configuration with k instantons, where λ_i are constants that correspond to the *size* of the instanton at the point y_i.

In a certain sense, SU(2)$-$instantons are also the 'building blocks' for instantons with general structure group. More precisely, let G be a compact semisimple Lie group, and let $\rho\colon \mathfrak{su}(2) \to \mathfrak{g}$ be any injective Lie algebra homomorphism. Then, for example,

$$\rho(A_{0,1}) = \frac{1}{|x|^2 + 1}\rho(\mathrm{Im}(q\,\mathrm{d}\overline{q}))$$

indeed defines a G−instanton on $\mathbb{R}^4$. While this guarantees the existence of G−instantons on the Euclidean space $\mathbb{R}^4$, observe that this instanton might be reducible (e.g. ρ can simply be the obvious inclusion of $\mathfrak{su}(2)$ into $\mathfrak{su}(r)$ for some $r \geqslant 3$) and that its charge depends on the choice of representation ρ. Furthermore, it is not clear whether every G−instanton can be obtained in this way.

Remark 1.64 (ADHM construction). For each $k \in \mathbb{N}$, the so-called 'ADHM construction', due to Atiyah et al. Atiyah et al. (1978), gives a correspondence between gauge-equivalence classes of ASD instantons ∇ with group SU(r) and fixed topological charge $\kappa(\nabla) = k$, and equivalence classes of certain systems of finite-dimensional algebraic data, for group SU(r) and index k Donaldson and Kronheimer (1990, §3.3). This gives a complete description of finite-energy ASD instantons over $\mathbb{R}^4$ with gauge group SU(r). $\lozenge$

Holomorphic structures and connections. We recall very briefly the Nirenberg–Newlander integrability theorem, relating holomorphic structures and certain types of connections on complex vector bundles over complex manifolds. In particular, this will serve as background material for the final paragraph §1.5 on ASD instantons and holomorphic structures.

Notation: We adopt the following conventions in this paragraph and the next one:

- Z: complex manifold of complex dimension m, i.e. a smooth $2m$−manifold endowed with an integrable almost complex structure J;

- $E \to Z$: (smooth) *complex* vector bundle over Z;

- $\Omega^{p,q}(Z, E) := \Gamma(\Lambda^{p,q}T^*Z_{\mathbb{C}} \otimes E)$: (p,q)−forms on Z with values on E;

- $\Omega^k(Z, E) = \bigoplus_{p+q=k} \Omega^{p,q}(Z, E)$: complex k−forms on Z with values on E.

Definition 1.65. A **holomorphic structure** $\mathcal{E}$ on a complex vector bundle $\pi\colon E \to Z$ is an additional complex manifold structure on the total space of E in such a way that π is a holomorphic map and the bundle admits an atlas of biholomorphic trivialisations. We call E endowed with the holomorphic structure $\mathcal{E}$ a **holomorphic vector bundle**, and we denote it by $\mathcal{E} \to Z$.

Alternatively, a holomorphic vector bundle $\mathcal{E} \to Z$ is a complex vector bundle $E \to Z$ determined by a $GL(r, \mathbb{C})$–cocycle $\{g_{\alpha\beta}\}$ of *holomorphic* transition functions $g_{\alpha\beta}\colon U_\alpha \cap U_\beta \to GL(r, \mathbb{C})$.

Given a holomorphic structure $\mathcal{E}$ on $E \to Z$, we can associate a unique $\mathbb{C}$–linear operator

$$\bar{\partial}_\mathcal{E}\colon \Omega^0(Z, E) \to \Omega^{0,1}(Z, E)$$

such that, for each $f \in C^\infty(Z, \mathbb{C})$ and $s \in \Gamma(E)$, we have

(i) $\bar{\partial}_\mathcal{E}(fs) = (\bar{\partial}f) \otimes s + f(\bar{\partial}_\mathcal{E}s)$;

(ii) If $U \subseteq Z$ is an open subset, then $\left(\bar{\partial}_\mathcal{E}s\right)|_U = 0$ if, and only if, s is a holomorphic map over U.

The construction of $\bar{\partial}_\mathcal{E}$ is as follows. By assumption, E admits an atlas of local frames $\{(e_{\alpha,1}, \dots, e_{\alpha,r})\}_\alpha$, whose associated transition functions $\{g_{\alpha\beta}\}$ are *holomorphic* maps. Given $s \in \Gamma(E)$, write

$$s|_{U_\alpha} = \sum_i s_\alpha^i \otimes e_{\alpha,i}, \quad s_\alpha^i \in C^\infty(U_\alpha, \mathbb{C}), i = 1, \dots, k.$$

and define

$$\left(\bar{\partial}_\mathcal{E}s\right)|_{U_\alpha} := \sum_i (\bar{\partial}s_\alpha^i) \otimes e_{\alpha,i}. \tag{1.66}$$

This operator clearly satisfies (i) and (ii). To see that it is well-defined, note that

$$\bar{\partial}(gv) = (\bar{\partial}g)v + g(\bar{\partial}v) = g(\bar{\partial}v),$$

whenever g is a *holomorphic* change of coordinates and v is the local representation of a section of E.

Of course, such operator $\bar{\partial}_\mathcal{E}$ can be extended to give rise to $\mathbb{C}$–linear operators

$$\bar{\partial}_\mathcal{E}\colon \Omega^{p,q}(Z, E) \to \Omega^{p,q+1}(Z, E), \quad \text{for all } p, q \geqslant 0,$$

such that

$$\bar{\partial}_\mathcal{E}(\omega \wedge s) = (\bar{\partial}\omega) \otimes s + (-1)^{p+q}\omega \wedge (\bar{\partial}_\mathcal{E}s),$$

whenever $\omega \in \Omega^{p,q}(Z)$ and $s \in \Omega^0(Z, E)$. Since Z is a complex manifold (therefore $\bar{\partial}^2 = 0$), it follows from the definition of $\bar{\partial}_{\mathcal{E}}$ (1.66) that $\bar{\partial}_{\mathcal{E}}^2 := \bar{\partial}_{\mathcal{E}} \circ \bar{\partial}_{\mathcal{E}} = 0$.

Now let ∇ be a (smooth) connection on the complex vector bundle $E \to Z$. Here we regard ∇ as a map from $\Gamma(E) = \Omega^0(Z, E)$ to $\Omega^1(Z, E)$ (see Remark 1.7). Then, the bi-degree splitting of $\Omega^1(Z, E)$ induces a corresponding splitting of ∇ as

$$\nabla = \partial_\nabla \oplus \bar{\partial}_\nabla : \Omega^0(Z, E) \to \Omega^{1,0}(Z, E) \oplus \Omega^{0,1}(Z, E).$$

The $\mathbb{C}$–linear operator $\bar{\partial}_\nabla : \Omega^0(Z, E) \to \Omega^{0,1}(Z, E)$ automatically satisfies (i), by the Leibniz rule:

$$\bar{\partial}_\nabla(fs) = \bar{\partial}f \otimes s + f\bar{\partial}_\nabla s,$$

for $f \in C^\infty(Z, \mathbb{C})$ and $s \in \Gamma(E)$.

Definition 1.67. A $\mathbb{C}$–linear operator $\bar{\partial}_E : \Omega^0(Z, E) \to \Omega^{0,1}(Z, E)$ is called a **partial connection** on E if it satisfies the '$\bar{\partial}$–Leibniz rule':

$$\bar{\partial}_E(fs) = \bar{\partial}f \otimes s + f\bar{\partial}_E s,$$

for each $f \in C^\infty(Z, \mathbb{C})$ and $s \in \Gamma(E)$.

Given a holomorphic structure $\mathcal{E}$ on E, it is clear that the induced operator $\bar{\partial}_{\mathcal{E}}$ is a partial connection. The non-trivial question is whether a given partial connection $\bar{\partial}_E$ comes from a holomorphic structure $\mathcal{E}$ on E, in the following sense:

Definition 1.68 (Integrability). A partial connection $\bar{\partial}_E$ on E is called **integrable** if it is equal to the partial connection $\bar{\partial}_{\mathcal{E}}$ induced by a holomorphic structure $\mathcal{E}$ on E.

In these terms, we quote the following crucial result from Donaldson and Kronheimer (1990, §2.2.2)):

Theorem 1.69 (Nirenberg–Newlander). *If $\bar{\partial}_E$ is a partial connection on E, then*

$$\bar{\partial}_E \text{ is integrable} \quad \Longleftrightarrow \quad \bar{\partial}_E^2 = 0.$$

If $\nabla = \partial_\nabla \oplus \bar{\partial}_\nabla$ is a connection on $E \to Z$, then

$$F_\nabla = \partial_\nabla^2 \oplus (\partial_\nabla\bar{\partial}_\nabla + \bar{\partial}_\nabla\partial_\nabla) \oplus \bar{\partial}_\nabla^2.$$

In particular,

$$F_\nabla^{0,2} = 0 \quad \Longleftrightarrow \quad \bar{\partial}_\nabla^2 = 0.$$

Definition 1.70 (Compatibility). A connection ∇ on $E \to Z$ is said to be **compatible** with a holomorphic structure $\mathcal{E}$ on E when $\bar{\partial}_\nabla$ is an integrable partial connection with $\bar{\partial}_\nabla = \bar{\partial}_\mathcal{E}$.

In conclusion, Theorem 1.69 implies the following relation between holomorphic structures and connections:

Corollary 1.71. *A connection ∇ on $E \to Z$ is compatible with a holomorphic structure $\mathcal{E}$ on E if, and only if, $F_\nabla^{0,2} = 0$.*

If, moreover, $E \to Z$ is *Hermitian*, a $\mathrm{U}(r)-$connection (*unitary* connection) is compatible with a holomorphic structure on E (i.e. $F_\nabla^{0,2} = 0$) if, and only if, $F_\nabla \in \Omega^{1,1}(Z, E)$. Indeed, if ∇ is unitary then $F_\nabla \in \Omega^2(Z, \mathfrak{u}(r)_E)$, hence

$$F_\nabla^{0,2} = -(F_\nabla^{2,0})^*.$$

On Hermitian bundles, a holomorphic structure distinguishes a unique compatible $\mathrm{U}(r)-$connection Donaldson and Kronheimer (1990, Lemma 2.1.54):

Proposition 1.72. *Suppose E is a $\mathrm{U}(r)-$bundle over Z. Then, a holomorphic structure $\mathcal{E}$ on E induces a unique compatible connection $\nabla \in \mathcal{A}(E)$, called the* **Chern connection** *of the holomorphic $\mathrm{U}(r)-$bundle $\mathcal{E} \to Z$.*

ASD instantons and holomorphic structures. To end this chapter, we now recall an important interpretation of the ASD instanton equation in the context of $\mathrm{SU}(r)-$bundles over complex Hermitian surfaces. The references are Donaldson and Kronheimer (ibid., pp. 46-47) and Scorpan (2005, pp. 369-370).

Let Z be a **Hermitian surface**, i.e. a (smooth) 4$-$manifold endowed with an integrable complex structure I and a Riemannian metric g with respect to which I is an orthogonal transformation. In particular, Z is a Riemannian 4$-$manifold with a preferred orientation fixed by I.

In this context, we have two decompositions of the complexified 2$-$forms of Z:

$$\Omega^2(Z) = \Omega^{2,0} \oplus \Omega^{1,1} \oplus \Omega^{0,2} \quad \text{and} \quad \Omega^2(Z) = \Omega_+^2 \oplus \Omega_-^2.$$

Denote by ω the fundamental 2-form of $(1,1)-$type induced by the pair (g, I):

$$\omega(X, Y) := g(IX, Y), \quad \forall X, Y \in \mathfrak{X}(Z).$$

Then ω induces a decomposition $\Omega^{1,1}(Z) = \Omega_0^{1,1} \oplus \Omega^0 \cdot \omega$, where $\Omega_0^{1,1} := (\Omega^0 \cdot \omega)^\perp \cap \Omega^{1,1}$.

By a straightforward local computation, the relation between the above decompositions is given by Donaldson and Kronheimer (1990, Lemma 2.1.57):

Proposition 1.73. *Let Z be a Hermitian complex surface as above. Then:*

- $\Omega^2_+ = \Omega^{2,0} \oplus \Omega^0 \cdot \omega \oplus \Omega^{0,2}.$

- $\Omega^2_- = \Omega^{1,1}_0.$

Therefore Donaldson and Kronheimer (ibid., Proposition 2.1.59):

Theorem 1.74. *Let $E \to Z$ be an* SU$(r)-$*bundle over a Hermitian surface Z. If $\nabla \in \mathcal{A}(E)$, then*

$$\nabla \text{ is an ASD instanton} \quad \Longleftrightarrow \quad \begin{cases} F^{0,2}_\nabla = 0 \text{ (integrability condition)} \\ \hat{F}_\nabla := F_\nabla \cdot \omega = 0 \end{cases}$$

Combining this result with the discussion of the last paragraph, one concludes that, in complex geometry, the ASD instanton condition splits naturally into two pieces, one of which has a simple geometric interpretation as an integrability condition. In particular, this suggests that ASD instantons, rather than SD instantons, are preferable in this setting. This is one of the reasons why one chooses to work with ASD, rather than SD, when doing gauge theory, even in the general context of oriented Riemannian 4−manifolds. From now on, we also adopt this perspective.

2

Instantons in higher dimensions

In the presence of suitable geometric structures on the manifold M^n, the 4–dimensional notion of instanton (cf. Section 1.5) can be generalised to *higher dimensional* contexts for $n > 4$. We present two approaches for such generalisation. The approach first explored by physicists Baulieu, Kanno, and Singer (1998) and Corrigan et al. (1983) is based on the presence of an appropriate $(n - 4)$–form on M. The second approach, originally introduced by Carrión (1998), one needs M to be equipped with an $N(H)$–structure, where $N(H)$ denotes the normaliser of some closed Lie subgroup $H \subseteq SO(n)$. These two points of view turn out to coincide in cases of interest, namely special holonomy manifolds, and were further popularized by the works of Donaldson and Thomas (1998), Tian (2000), Donaldson and Segal (2011) et al.

We begin with a discussion of Berger's classification theorem of Riemannian holonomy groups (Section 2.1). In particular, we give short descriptions of the special geometries associated to the holonomy groups $U(m)$ and $SU(m)$, respectively Kähler and Calabi–Yau, as well as G_2 and $Spin(7)$. Next, in Section 2.2, we introduce the language of calibrated geometry and its relations with special holonomy manifolds. Then, in terms of both aforementioned approaches, in Section 2.3 we explain generalisations of the notion of instanton for oriented Riemannian n–manifolds, $n \geqslant 4$, endowed with an appropriate geometric structure. In fact, we will be interested in those cases for which the holonomy group of g is realised as a normaliser $N(H) \subsetneq SO(n)$ appearing in Berger's list of special geometries. We pay particular attention to the corresponding notions of instanton associated to the holonomy reductions $\mathrm{SU}(m) = N(\mathrm{U}(m)) \subseteq \mathrm{SO}(2m)$, $\mathrm{G}_2 = N(\mathrm{G}_2) \subseteq \mathrm{SO}(7)$ and $\mathrm{Spin}(7) = N(\mathrm{Spin}(7)) \subseteq \mathrm{SO}(8)$, with an emphasis on the last two 'exceptional' cases.

2.1 Riemannian metrics with special holonomy group

The main references for this section are Bryant (1986) and Joyce (2000, 2006, 2007).

Riemannian holonomy groups and Berger's classification. Let (M, g) be a Riemannian $n-$manifold and denote by D^g its Levi-Civita connection on the real $\mathrm{O}(n)-$bundle TM. Recall that D^g is uniquely determined by the following properties Joyce (2000, Theorem 3.1.1):

(i) D^g is *torsion-free*, i.e. $D^g_X Y - D^g_Y X = [X, Y]$ for all $X, Y \in \mathfrak{X}(M)$;

(ii) D^g is compatible with g, i.e. $D^g g = 0$.

Write $\mathrm{Hol}_x(g) := \mathrm{Hol}_x(D^g)$ for the holonomy group of g at x (cf. Section 1.2). Since the subgroup of $\mathrm{GL}(T_x M)$ preserving $g|_{T_x M}$ is $\mathrm{O}(T_x M)$, the metric compatibility (ii) implies, via Theorem 1.32, that $\mathrm{Hol}_x(g) \subseteq \mathrm{O}(T_x M)$. In particular, we can regard $\mathrm{Hol}(g) := \mathrm{Hol}(D^g)$ as a subgroup of $O(n)$, well-defined up to conjugation in $O(n)$. By connectedness, the restricted holonomy group $\mathrm{Hol}^0(g) := \mathrm{Hol}^0(D^g)$ is a subgroup of $\mathrm{SO}(n)$, defined up to conjugation (by $\mathrm{O}(n)$), and the holonomy algebra $\mathfrak{hol}(g)$ is a Lie subalgebra of $\mathfrak{so}(n)$, defined up to the adjoint action (by $\mathrm{O}(n)$).

The Riemann curvature tensor $R^g := F_{D^g}$ of g has a number of symmetries, besides the obvious skew-symmetry in its first two arguments. To express such symmetries it is convenient to lower the last index of R^g:

$$Rm^g(X, Y, Z, W) := g(R^g(X, Y)Z, W), \quad \forall X, Y, Z, W \in \mathfrak{X}(M).$$

We shall refer to both R^g and Rm^g as the *Riemann curvature* of g. In terms of components, with respect to any local frame, the tensor R^g is represented by R^i_{jkl} and the tensor Rm^g is represented by R_{ijkl}. Also, we denote the total covariant derivative $D^g Rm^g$ in components by $R_{ijkl;m}$. The following result summarises important symmetries of Rm^g and $D^g Rm^g$ Joyce (2007, Theorem 3.1.2).

Proposition 2.1. *Let (M, g) be a Riemannian manifold with Riemann curvature R_{ijkl}. Then:*

$$R_{ijkl} = -R_{ijlk} = -R_{jikl} = R_{klij}, \tag{2.2}$$

$$R_{ijkl} + R_{jkil} + R_{kijl} = 0, \quad \text{(algebraic/}1^{st}\text{ Bianchi identity)} \tag{2.3}$$

$$R_{ijkl;m} + R_{ijlm;k} + R_{ijmk;l} = 0. \quad \text{(differential/}2^{nd}\text{ Bianchi identity)} \tag{2.4}$$

Remark 2.5. (2.4) is simply a rephrasing of the Bianchi identity (1.21) in this context. $\diamond$

At each point $x \in M$, regarding $\mathfrak{hol}_x(g)$ as a subspace of the anti-symmetric endomorphisms $\mathfrak{so}(T_x M)$ of $T_x M$, it follows from Proposition 1.38 that R^i_{jkl} lies in $\Lambda^2 T_x^* M \otimes \mathfrak{hol}_x(g)$. By equation (2.2), we see that R_{ijkl} is an element of $\Lambda^2 T_x^* M \otimes \Lambda^2 T_x^* M$, so that identifying $\mathfrak{so}(T_x M)$ with $\Lambda^2 T_x^* M$ using g, we can also think of R_{ijkl} as an element of $\Lambda^2 T_x^* M \otimes \mathfrak{hol}_x(g)$. Furthermore, using the first Bianchi identity (2.3), we get Joyce (ibid., Theorem 3.1.7):

Proposition 2.6. *Let (M, g) be a Riemannian manifold with Riemann curvature R_{ijkl}. Then R_{ijkl} lies in the subspace $S^2 \mathfrak{hol}_x(g)$ of $\Lambda^2 T_x^* M \otimes \Lambda^2 T_x^* M$ at each point $x \in M$.*

Together with the Bianchi identities of Proposition 2.1, this result gives quite strong restrictions on the curvature tensor of a Riemannian metric g with a prescribed holonomy group $\text{Hol}(g)$ Joyce (ibid., p. 43). Combined with the Ambrose–Singer theorem 1.39, this is the basis of the (algebraic) classification of Riemannian holonomy groups.

A theorem due to de Rham Joyce (ibid., Theorem 3.2.7)) shows that if (M, g) is a complete, simply-connected Riemannian manifold, then there exist complete, simply-connected Riemannian manifolds (M_j, g_j) for $j = 1, \ldots, k$, such that the holonomy representation of $\text{Hol}(g_j)$ is irreducible, (M, g) is isometric to the Riemannian product $(M_1 \times \ldots \times M_k, g_1 \times \ldots \times g_k)$, and $\text{Hol}(g) = \text{Hol}(g_1) \times \ldots \times \text{Hol}(g_k)$. Thus, in looking for a classification of the possible holonomy groups of (M^n, g), we are mainly interested in the cases where $\text{Hol}^0(g)$ acts irreducibly on $\mathbb{R}^n$.

In 1955, Berger (1955) gave a list of all the possible irreducible holonomy groups for Riemannian metrics Bryant (1986):

Theorem 2.7 (Berger). *Let M be a connected, simply-connected[1] n-dimensional manifold, and let g be a Riemannian metric on M. Suppose that, for some $x \in M$, $\text{Hol}_x(g)$ acts irreducibly on $T_x M$. Then either g is a locally symmetric metric or else one of the following holds:*

---

[1] If $\pi(M) \neq 1$ then the universal cover $(\tilde{M}, \tilde{g})$ of (M, g) has $\text{Hol}(\tilde{g}) = \text{Hol}^0(g)$.

(i) $\mathrm{Hol}(g) = \mathrm{SO}(n)$,

(ii) $n = 2m$, $m \geqslant 2$ *and* $\mathrm{Hol}(g) = \mathrm{U}(m)$,

(iii) $n = 2m$, $m \geqslant 2$ *and* $\mathrm{Hol}(g) = \mathrm{SU}(m)$,

(iv) $n = 4m$, $m \geqslant 2$ *and* $\mathrm{Hol}(g) = \mathrm{Sp}(m)$,

(v) $n = 4m$, $m \geqslant 2$ *and* $\mathrm{Hol}(g) = \mathrm{Sp}(m) \cdot \mathrm{Sp}(1)$,

(vi) $n = 7$ *and* $\mathrm{Hol}(g) = \mathrm{G}_2$,

(vii) $n = 8$ *and* $\mathrm{Hol}(g) = \mathrm{Spin}(7)$.

Remark 2.8. A Riemannian metric g on M is called *locally symmetric* if every point $p \in M$ admits an open neighbourhood U_p in M, and an involutive isometry $\sigma_p \colon U_p \to U_p$ with unique fixed point p. For more on this we refer the reader to Joyce (2007, §3.3). $\qquad\qquad\qquad\qquad\qquad\qquad\qquad\qquad\qquad\diamond$

Remark 2.9. Later on, Simons (1962) gave another proof of Theorem 2.7. See also the more recent proof by Olmos (2005). $\qquad\qquad\qquad\qquad\qquad\qquad\diamond$

From now on we shall refer to the list of groups (i)-(vii) as *Berger's list*. A very thorough discussion of Berger's theorem, including discussions of each of the geometries associated to the groups in Berger's list, analogies with the four normed division algebras, and the principles behind Berger's original proof, can be found in Joyce's book Joyce (2007, §3.4).

It can be shown that the space of Riemannian metrics g on M^n for which $\mathrm{Hol}(g) = \mathrm{SO}(n)$ is both open and dense in the space of Riemannian metrics on M. Thus, one says that $\mathrm{SO}(n)$ is the holonomy group of a *generic* metric on M. The other groups on Berger's list are called **special holonomy groups**. In what follows, we give brief descriptions of metrics with these holonomy groups, except the cases (iv) and (v) which will not be fundamental for our later purposes.

Metrics with holonomy $\mathbf{U}(m) \subseteq \mathbf{O}(2m)$.
(cf. S. Salamon (1989, Chapter 3))

Let $(z^1, \ldots, z^m)$ be complex coordinates on $\mathbb{C}^m$. The unitary group $\mathrm{U}(m)$ may be defined as the set of complex linear endomorphisms of $\mathbb{C}^m$ preserving the Hermitian form

$$\eta_0 = \sum_{j=1}^{m} \mathrm{d}z^j \otimes \mathrm{d}\overline{z^j}.$$

Defining real coordinates $(x^1, \ldots, x^{2m})$ on $\mathbb{C}^m \simeq \mathbb{R}^{2m}$ by

$$z^j = x^{2j-1} + \mathbf{i}x^{2j}, \quad j = 1, \ldots, m,$$

we can write

$$\eta_0 = \sum_{j=1}^{m} \mathrm{d}x^j \otimes \mathrm{d}x^j - 2\mathbf{i} \sum_{j=1}^{m} \mathrm{d}x^{2j-1} \wedge \mathrm{d}x^{2j}.$$

The real part $g_0 = \mathrm{Re}(\eta_0)$ is the standard Euclidean inner product on $\mathbb{R}^{2m}$, so that $\mathrm{U}(m)$ acts on $\mathbb{R}^{2m}$ as the subgroup of $\mathrm{O}(2m)$ which fixes the real 2–form $-2\omega_0 := \mathrm{Im}(\eta_0)$. The group $\mathrm{U}(m)$ also commutes with the real endomorphism I_0 of $\mathbb{R}^{2m}$ such that $I_0 dx^{2j-1} = \mathrm{d}x^{2j}$ and $I_0 \mathrm{d}x^{2j} = -\mathrm{d}x^{2j-1}$ ($j = 1, \ldots, m$). It can be shown that ω_0 and I_0 are equivalent in the presence of the inner product g_0; for instance, $\omega_0(x, y) = g_0(I_0 x, y)$ for all $x, y \in \mathbb{R}^{2m}$.

It follows from the holonomy principle (Theorem 1.32) that a Riemannian metric g on a $2m$–dimensional manifold Z^{2m} has holonomy $\mathrm{Hol}(g) \subseteq \mathrm{U}(m)$ if, and only if, Z admits natural tensors $I \in \mathrm{End}(TZ)$ and $\omega \in \Omega^2(Z)$, parallel with respect to the Levi-Civita connection D^g, such that g, I, and ω can be written in the form g_0, I_0, and ω_0 at each point of Z. A Riemannian metric g on a $2m$–dimensional manifold Z^{2m} with $\mathrm{Hol}(g) \subseteq U(m)$ is called a *Kähler metric*.

A $\mathrm{U}(m)$–structure on a smooth $2m$–manifold Z is specified by a pair (I, ω), where $I \in \mathrm{End}(TZ)$ is an almost complex structure, $I^2 = -\mathbb{1}$, and $\omega \in \Omega^2(Z)$ is a non-degenerate real 2–form such that $g(\cdot, \cdot) := \omega(\cdot, I \cdot)$ defines a Riemannian metric on Z. A $\mathrm{U}(m)$–structure (I, ω) on Z^{2m} is *torsion free* when both I and ω are D^g–parallel with respect to the induced metric g. A $2m$–dimensional manifold Z^{2m} endowed with a torsion-free $U(m)$–structure (I, ω) is called a *Kähler m–fold* and ω its *Kähler form*. The following facts are standard Joyce (2000, §4.4):

Proposition 2.10. *Let (I, ω) be a $U(m)$–structure on Z^{2m}. Denote by g the natural Riemannian metric induced by (I, ω). Then the following are equivalent:*

(i) g is a Kähler metric.

(ii) (I, ω) is torsion-free.

(iii) I is integrable[2] and $\mathrm{d}\omega = 0$.

Note that Kähler m–folds are essentially Riemannian $2m$–manifolds with holonomy contained in $U(m)$. Henceforth, we will denote a Kähler m–fold by a pair (Z^{2m}, ω), omitting the underlying complex structure I and metric g.

---

[2]i.e. I is induced from a complex manifold structure on Z

Metrics with holonomy $SU(m) \subseteq O(2m)$.
(cf. Corti et al. (2013, §2))

As before, we identify $\mathbb{R}^{2m} \simeq \mathbb{C}^m$, with complex coordinates $(z^1, \ldots, z^m)$, and define a complex m−form Υ_0 on $\mathbb{C}^m$ by

$$\Upsilon_0 := dz^1 \wedge \ldots \wedge dz^m.$$

The subgroup of $U(m) \subseteq O(2m)$ preserving g_0, ω_0 and Υ_0 is $SU(m)$.

By the holonomy principle, a Riemannian metric g on a $2m$−dimensional manifold Z^{2m} has holonomy $\mathrm{Hol}(g) \subseteq SU(m)$ if, and only if, g is a Kähler metric, say, with associated complex structure I and Kähler form ω, and further Z admits a natural D^g−parallel complex $(m, 0)$−form Υ such that g, I, ω and Υ have pointwise models g_0, I_0, ω_0 and Υ_0. A Riemannian metric g on a $2m$−dimensional manifold Z^{2m} with $\mathrm{Hol}(g) \subseteq SU(m)$ is called a *Calabi–Yau metric*.

An $SU(m)$−*structure* on a smooth $2m$−manifold Z is specified by a triple (I, ω, Υ), where (I, ω) defines a $U(m)$−structure on Z, and Υ is a nowhere vanishing complex $(m, 0)$−form on (Z, I) satisfying

$$\frac{\omega^m}{m!} = \mathbf{i}^{m^2} 2^{-m} \Upsilon \wedge \overline{\Upsilon}. \tag{2.11}$$

This is a normalisation condition that the natural volume forms induced by ω and Υ are equal, or equivalently that $|\Upsilon|^2 = 2^m$ with respect to the induced metric[3]. An $SU(m)$−structure (I, ω, Υ) on Z^{2m} is called *torsion-free* when $D^g \Upsilon = D^g \omega = 0$ with respect to its induced metric g. A $2m$−manifold Z^{2m} endowed with a torsion-free $SU(m)$−structure (I, ω, Υ) is called a *Calabi–Yau* m−fold.

One can show that given an $SU(m)$−structure (I, ω, Υ) on Z^{2m}, then $d\Upsilon = 0$ implies that the complex structure I is integrable and Υ is a holomorphic $(m, 0)$−form. In particular, Υ holomorphically trivialises the canonical bundle $K_Z = \Lambda^m (T^{1,0}Z)^*$ of (Z, I). Since the first Chern class $c_1(Z) := c_1(T^{1,0}Z)$ turns out to be a characteristic class of K_Z, namely $-c_1(K_Z)$, it follows that $c_1(Z) = 0$. If also $d\omega = 0$ then Z is a Kähler manifold, so that the induced metric g has $\mathrm{Hol}(g) \subseteq U(m)$. Furthermore, since Υ is a holomorphic form of constant norm, the later condition forces $D^g \Upsilon = 0$, so that the holonomy of g reduces further to $\mathrm{Hol}(g) \subseteq SU(m)$. In particular, an $SU(m)$−structure (I, ω, Υ) is torsion-free if, and only if, $d\Upsilon = d\omega = 0$.

The well-known linear relation between the curvature of the canonical bundle and the Ricci curvature of a Kähler metric implies Joyce (2007, Proposition 7.1.1):

---

[3]This implies that $\mathrm{Re}(\Upsilon)$ has comass $\leqslant 1$ (cf. Section 2.2.2).

Proposition 2.12. *Suppose (Z^{2m}, ω) is a Kähler m−fold and let g be its associated compatible Riemannian metric. Then $\mathrm{Hol}^0(g) \subseteq \mathrm{SU}(m)$ if, and only if, g is Ricci-flat ($\mathrm{Ric}^g \equiv 0$).*

Finally, a fundamental result in this context is Yau's solution of the Calabi conjecture Yau (1978), which has the following important consequence Joyce (2007, Theorem 7.1.2):

Theorem 2.13. *Let (Z^{2m}, I) be a compact complex manifold admitting some Kähler metric and such that $c_1(Z) = 0$. Then there is a unique Ricci-flat Kähler metric in the cohomology class of each Kähler form on Z.*

Since a generic Kähler metric on a complex m−fold has holonomy $\mathrm{U}(m)$, in the light of Proposition 2.12 we see the above theorem constructs metrics with special holonomy $\subseteq \mathrm{SU}(m)$ on compact complex m−folds.

Henceforth, we will denote a Calabi–Yau m−fold by a triple (Z, ω, Υ), omitting the underlying complex structure I and metric g.

The exceptional cases $G_2 \subseteq \mathrm{SO}(7)$ and $\mathrm{Spin}(7) \subseteq \mathrm{SO}(8)$.
(cf. Bryant (1986, 1987) and Joyce (2007, Chapter 11))

We start with a definition of the Lie group G_2 due to Bryant [ibid.]:

Definition 2.14. Let $(x^1, \ldots, x^7)$ be Euclidean coordinates on $\mathbb{R}^7$. Define a 3−form ϕ_0 on $\mathbb{R}^7$ by

$$\phi_0 := \mathrm{d}x^{123} - \mathrm{d}x^{145} - \mathrm{d}x^{167} - \mathrm{d}x^{246} + \mathrm{d}x^{257} - \mathrm{d}x^{347} - \mathrm{d}x^{356}. \quad (2.15)$$

Here we write $\mathrm{d}x^{ij\cdots l}$ as shorthand for $\mathrm{d}x^i \wedge \mathrm{d}x^j \wedge \ldots \wedge \mathrm{d}x^l$. The subgroup of $\mathrm{GL}(7, \mathbb{R})$ preserving ϕ_0 under the standard (pull-back) action is the exceptional Lie group G_2:

$$G_2 := \{g \in \mathrm{GL}(7, \mathbb{R}) : g^*\phi_0 = \phi_0\}.$$

Remark 2.16. Our definition of ϕ_0 differs from the one given by Bryant by an orientation-preserving change of coordinates. Our sign conventions follows D. A. Salamon and Walpuski (2017) and Walpuski (2013b).

A useful way to interpret ϕ_0 is to write $\mathbb{R}^7 \simeq \mathbb{R}^3 \oplus \mathbb{R}^4$, with respective coordinates (x^1, x^2, x^3) and (x^4, x^5, x^6, x^7), and the standard choice of orientations

$$\mathrm{vol}_3 := \mathrm{d}x^{123} \quad \text{and} \quad \mathrm{vol}_4 := \mathrm{d}x^{4567},$$

on $\mathbb{R}^3$ and $\mathbb{R}^4$, respectively. Note that the 2−forms

$$\eta_1^+ := \mathrm{d}x^{45} + \mathrm{d}x^{67},$$
$$\eta_2^+ := \mathrm{d}x^{46} + \mathrm{d}x^{75},$$
$$\eta_3^+ := \mathrm{d}x^{47} + \mathrm{d}x^{56},$$

give us an orthogonal basis for the selfdual 2−forms on $\mathbb{R}^4$. With these identifications, we can write

$$\phi_0 = \mathrm{vol}_3 - \mathrm{d}x^1 \wedge \eta_1^+ - \mathrm{d}x^2 \wedge \eta_2^+ - \mathrm{d}x^3 \wedge \eta_3^+.$$

$$\diamondsuit$$

By definition, G_2 is a closed Lie subgroup of $GL(7, \mathbb{R})$. Moreover, one can check directly that for every $x, y \in \mathbb{R}^7$, we have

$$(x \lrcorner \phi_0) \wedge (y \lrcorner \phi_0) \wedge \phi_0 = 6g_7(x, y)\mathrm{vol}_7, \qquad (2.17)$$

where g_7 and vol_7 denotes, respectively, the standard metric and orientation of $\mathbb{R}^7$. In particular, we see that $G_2 \subseteq SO(7)$.

The following theorem summarises some general facts about the Lie group G_2 Bryant (1987, Theorem 1, p. 539).

Theorem 2.18. G_2 *is a* $14-$*dimensional compact,* $2-$*connected, simple Lie group.*

Definition 2.19. Let V be a $7-$dimensional real vector space. A $3-$form $\phi \in \Lambda^3 V^*$ is said to be **positive** if there exists a linear isomorphism $u \colon V \to \mathbb{R}^7$ so that $\phi = u^*\phi_0$, where $\phi_0 \in \Lambda^3(\mathbb{R}^7)^*$ is given by (2.15). The set of positive $3-$forms on V is denoted by $\Lambda_+^3 V^*$.

Remark 2.20. Note that $\Lambda_+^3 V^* \simeq GL(7, \mathbb{R})/G_2$, so straightforward dimension counting shows that $\Lambda_+^3 V^* \subset \Lambda^3 V^*$ is an open subset. $\qquad \diamondsuit$

Definition 2.21. Let Y^7 be a smooth manifold and denote by $\Lambda_+^3(T^*Y)$ be the (open) subbundle of $\Lambda^3 T^*Y$ whose fibre over $y \in Y$ is $\Lambda_+^3(T_y^*Y)$ and by $\Omega_+^3(Y)$ its space of smooth sections. An element $\phi \in \Omega_+^3(Y)$ is called a **positive** $3-$**form** on Y.

By the holonomy principle (Theorem 1.32), it follows that a (connected) Riemannian manifold (Y^7, g) has $\mathrm{Hol}(g) \subseteq G_2$ if, and only if, Y possesses a parallel positive $3-$form $\phi \in \Omega_+^3(Y)$.

Note that a positive $3-$form on Y^7 is equivalent to a G_2-structure on (the frame bundle $\mathcal{F}$ of) Y^7. Indeed, given $\phi \in \Omega_+^3(Y)$ we can form

$$\mathcal{P}_\phi := \{u \in \mathcal{F} : u^*\phi_0 = \phi_y, \text{ where } u \colon T_y Y \to \mathbb{R}^7\}.$$

It is easy to see $\mathcal{P}_\phi$ defines a principal subbundle of $\mathcal{F}$ with fibre G_2, i.e. a G_2-structure on Y. Conversely, a G_2-structure $\mathcal{P} \subseteq \mathcal{F}$ on Y determines a unique positive $3-$form $\phi \in \Omega_+^3(Y)$ by

$$\phi_y := u_y^*\phi_0,$$

where $u_y \in \mathcal{P}_y$, for all $y \in Y$. This is well-defined precisely because $\mathcal{P}$ is a principal G_2–subbundle: two frames $u_y, u'_y \in \mathcal{P}_y$ are related as $u'_y = g^{-1} \circ u_y$, for some $g \in G_2 = \mathrm{Stab}(\phi_0)$. It is clear that such constructions are inverse of each other. Henceforth we will *not distinguish* between G_2–structures and positive 3–forms on Y^7.

Since $G_2 \subseteq SO(7)$, a G_2–structure ϕ on Y^7 determines a Riemannian metric g_ϕ and an orientation vol_ϕ on Y. Indeed, these are uniquely determined pointwise by the relation (2.17). In particular, ϕ determines a $*$–Hodge operator on $\Lambda^\bullet T^* Y$.

Definition 2.22. A G_2–structure ϕ on Y^7 is called **torsion-free** when it is parallel with respect to the induced Levi-Civita connection:

$$D^{g_\phi}\phi = 0. \tag{2.23}$$

If ϕ is a torsion-free G_2–structure on Y^7, the pair (Y^7, ϕ) is called a G_2–**manifold**.

Thus, a G_2–manifold (Y^7, ϕ) is essentially a Riemannian manifold (Y^7, g_ϕ) with $\mathrm{Hol}(g_\phi) \subseteq G_2$.

Remark 2.24. The torsion-free condition (2.23) turns out to be a very complicated *non-linear* p.d.e. on ϕ. The non-linearity is due to the dependency of the metric g_ϕ itself (hence the Levi-Civita connection) on ϕ. $\diamond$

Example 2.25. $(\mathbb{R}^7, \phi_0)$, with ϕ_0 given by (2.15), is the model example of G_2–manifold.

The following theorem from Fernández and Gray (1982, Theorem 5.2) (see also S. Salamon (1989, Lemma 11.5, p. 160)) gives a non-trivial characterisation for the torsion-free condition (2.23):

Theorem 2.26 (Fernández–Gray). *Let Y^7 be a connected manifold and let $\phi \in \Omega^3_+(Y)$. Denote by $*$ the Hodge star operator induced by ϕ on Y. Then the following are equivalent:*

(i) (Y, ϕ) is a G_2–manifold.

(ii) $\mathrm{d}\phi = 0 = \mathrm{d} * \phi$.

Remark 2.27. Again, since $*$ depends on ϕ, $\mathrm{d} * \phi = 0$ is a non-linear condition on ϕ. $\diamond$

Exploring curvature restrictions imposed by the holonomy group, just as in Theorem 2.6, and using some representation theory, one can prove the following S. Salamon (ibid., Proposition 11.8):

Proposition 2.28. *If g is a Riemannian metric on a (connected) 7−manifold Y^7 with $\mathrm{Hol}(g) \subseteq G_2$, then g is Ricci-flat ($\mathrm{Ric}^g \equiv 0$).*

Moreover, from the classification of Riemannian holonomy groups (Theorem 2.7), one has Joyce (2007, Theorem 11.1.7):

Theorem 2.29. *The only non-trivial connected Lie subgroups of G_2 which can occur as holonomy of a Riemannian 7−manifold are:*

(i) $\mathrm{SU}(2)$*, acting on* $\mathbb{R}^7 \simeq \mathbb{R}^3 \oplus \mathbb{C}^2$*, trivial on* $\mathbb{R}^3$*, standard on* $\mathbb{C}^2$*,*

(ii) $\mathrm{SU}(3)$*, action on* $\mathbb{R}^7 \simeq \mathbb{R} \oplus \mathbb{C}^3$*, trivial on* $\mathbb{R}$*, standard on* $\mathbb{C}^3$*.*

Thus, if ϕ is torsion-free G_2−structure on a 7−manifold, then $\mathrm{Hol}^0(g_\phi)$ is one of $\{1\}$, $\mathrm{SU}(2)$, $\mathrm{SU}(3)$ or G_2.

This theorem implies that we can obtain G_2−manifolds from certain lower dimensional geometries. More precisely, the inclusions $\mathrm{SU}(2) \subseteq G_2$ and $\mathrm{SU}(3) \subseteq G_2$ imply that from each Calabi–Yau 2− or 3−fold we can make a G_2−manifold.

Example 2.30 (G_2−manifolds from Calabi–Yau 2−folds). Let (Z^4, ω, Υ) be a Calabi–Yau 2−fold, and let (x^1, x^2, x^3) be coordinates on $\mathbb{R}^3$ or $T^3 := S^1 \times S^1 \times S^1$. Then the 3−form

$$\phi := \mathrm{d}x^{123} - \mathrm{d}x^1 \wedge \omega - \mathrm{d}x^2 \wedge \mathrm{Re}(\Upsilon) - \mathrm{d}x^3 \wedge \mathrm{Im}(\Upsilon)$$

defines a torsion-free G_2−structure on $Y^7 := \mathbb{R}^3 \times Z^4$ or $T^3 \times Z^4$ compatible with the natural product metric and orientation structures.

Example 2.31 (G_2−manifolds from Calabi–Yau 3−folds). Let (Z^6, ω, Υ) be a Calabi–Yau 3−fold. Let t be a coordinate on $\mathbb{R}$ or S^1. Then the 3−form

$$\phi := \mathrm{d}t \wedge \omega + \mathrm{Re}(\Upsilon)$$

defines a torsion-free G_2−structure on $Y^7 := \mathbb{R} \times Z^6$ or $S^1 \times Z^6$, compatible with the natural product metric and orientation structures.

Note that the above examples have holonomy strictly contained in G_2. Examples of metrics with holonomy *exactly* G_2 are much harder to come by. In fact, for almost three decades after Berger's classification (Theorem 2.7), the exceptional holonomy groups G_2 and $\mathrm{Spin}(7)$ posed a mystery, as to whether examples existed at all. Eventually, Bryant (1987) proved the local existence of such metrics, and constructed some explicit incomplete examples. Then, Bryant–Salamon Bryant and S. Salamon (1989) constructed the first examples of *complete* metrics with holonomy (exactly) G_2 and Spin(7) on noncompact manifolds. Later, Joyce (1996) constructed the first examples of metrics with holonomy (exactly) G_2 and Spin(7) on *compact* manifolds; also see Joyce (2000).

Another particularly important method in the construction of compact G_2-manifolds, with full holonomy G_2, is the so-called *twisted connected sum* construction. From a pair of smooth asymptotically cylindrical Calabi–Yau 3$-$folds $V_\pm$ Haskins, Hein, and Nordström (2015), there is a non-trivial way to glue the products $S^1 \times V_\pm$, truncated sufficiently far along one tubular end, so as to produce a compact G_2-manifold $Y^7 := \left(S^1 \times V_+\right) \widetilde{\#} \left(S^1 \times V_-\right)$ with holomomy exactly G_2. This method was first developed by Kovalev (2003), based on an insight by Donaldson. Then the construction was improved by Kovalev and Lee (2011) and, more recently, corrected and extended significantly by Corti et al. (2015). The twisted connected sum construction provided a major breakthrough in the study of G_2-manifolds, allowing for hundreds of thousands of diffeomorphism types of examples to be mass-produced.

At the time of writing, there are only two other methods for the production of compact manifolds with holonomy exactly G_2, by Joyce (2000, §§11, 12) and Joyce–Karigiannis Joyce and Karigiannis (2017). Both are based on orbifold resolution techniques yielding G_2-structures with *small torsion*, which can then be perturbed into a torsion-free solution by an elliptic p.d.e. argument.

We now turn to a brief discussion of the holonomy group Spin(7) (cf. Joyce (2000, §10.5)).

Definition 2.32. Define a 4$-$form Φ_0 on $\mathbb{R}^8 = \mathbb{R} \times \mathbb{R}^7$, in coordinates $(x^0, x^1, \ldots, x^7)$, by

$$\Phi_0 := \mathrm{d}x^0 \wedge \phi_0 + \psi_0, \tag{2.33}$$

where ϕ_0 is the 3$-$form (2.15), and $\psi_0 := *_7\phi_0$. The GL$(8, \mathbb{R})-$stabiliser of Φ_0 under the (standard) pull-back action is the Lie group Spin(7):

$$\mathrm{Spin}(7) := \{g \in \mathrm{GL}(8, \mathbb{R}) : g^*\Phi_0 = \Phi_0\}.$$

Note that Φ_0 is manifestly selfdual with respect to the Euclidean metric on $\mathbb{R}^8$.

Theorem 2.34 (Bryant (1987, Theorem 4)). Spin(7) *is a simple, compact and* 1$-$*connected Lie group of dimension 21. Furthermore,* Spin(7) *is a subgroup of* SO(8).

Definition 2.35. Let W^8 be a 8$-$dimensional real vector space. A 4$-$form $\Phi \in \Lambda^4 W^*$ is called **definite** if there exists a linear isomorphism $u : W \to \mathbb{R}^8$

such that $\Phi = u^*\Phi_0$. We denote by $\Lambda^4_+(W^*)$ the set of definite 4−forms on W.

Let X^8 be a smooth manifold. Define the bundle $\Lambda^4_+(T^*X)$ of definite 4−forms on X to be the subbundle of $\Lambda^4 T^*X$ whose fibre at $x \in X$ is $\Lambda^4_+(T^*_x X)$. A smooth section $\Phi \in \Gamma(\Lambda^4_+(T^*X))$ is called a **definite** 4−**form** on X. The space of definite 4−forms on X is denoted by $\Omega^4_+(X)$.

Note that a definite 4−form $\Phi \in \Omega^4_+(X)$ determines and is determined by a unique Spin(7)−structure on X. Thus, it is customary to call such Φ a Spin(7)−structure on X.

Since Spin(7) $\subseteq$ SO(8), a Spin(7)−structure Φ on X determines a unique Riemannian metric g_Φ and orientation vol_Φ on X; in particular, we have an associated $*$−operator acting on $\Lambda^\bullet T^*X$.

Definition 2.36. A Spin(7)−structure Φ on a smooth 8−manifold X is called **torsion free** if $D^{g_\Phi}\Phi = 0$. A pair (X^8, Φ) where Φ is a torsion-free Spin(7)−structure on X^8 is called a Spin(7)−**manifold**.

By the holonomy principle (Theorem 1.32), a connected Riemannian 8−manifold (X^8, g) has $\mathrm{Hol}(g) \subseteq$ Spin(7) if, and only if, (X, g) has a torsion-free Spin(7)−structure Φ. Thus, a Spin(7)−manifold (X^8, Φ) is essentially a Riemannian 8−manifold (X^8, g_Φ) with $\mathrm{Hol}(g_\Phi) \subseteq$ Spin(7).

Example 2.37. $(\mathbb{R}^8, \Phi_0)$ where Φ_0 is given by (2.33) is the model example.

The next results are analogues of Theorem 2.26, Proposition 2.28 and Theorem 2.29.

Theorem 2.38 (S. Salamon (1989, Lemma 12.4)). *Let Φ be a* Spin(7)−*structure on X^8. Then the following are equivalent:*

(i) (X, Φ) is a Spin(7)−*manifold.*

(ii) $\mathrm{d}\Phi = 0$.

Proposition 2.39 (S. Salamon (ibid., Corollary 12.6)). *Let (X^8, g) be a connected Riemannian manifold, with* $\mathrm{Hol}(g) \subseteq$ Spin(7). *Then g is Ricci-flat.*

From Berger's classification theorem (Theorem 2.7), one deduces:

Theorem 2.40 (Joyce (2007, Theorem 11.4.7)). *The only* non-trivial *connected Lie subgroups of* Spin(7) *which can be holonomy groups of Riemannian metrics on 8−manifolds are:*

(i) SU(2), *acting on* $\mathbb{R}^8 \simeq \mathbb{R}^4 \oplus \mathbb{C}^2$, *trivial on* $\mathbb{R}^4$ *and standard on* $\mathbb{C}^2$;

(ii) SU(2) × SU(2), *acting on* $\mathbb{R}^8 \simeq \mathbb{C}^2 \oplus \mathbb{C}^2$, *in the obvious way;*

(iii) SU(3), *acting on* $\mathbb{R}^8 \simeq \mathbb{R}^2 \oplus \mathbb{C}^3$, *trivial on* $\mathbb{R}^2$ *and standard on* $\mathbb{C}^3$;

(iv) G_2, *acting on* $\mathbb{R}^8 \simeq \mathbb{R} \oplus \mathbb{R}^7$, *trivial on* $\mathbb{R}$ *and standard on* $\mathbb{R}^7$.

(v) Sp(2), *acting as usual on* $\mathbb{R}^8 \simeq \mathbb{H}^2$.

(vi) SU(4), *acting as usual on* $\mathbb{R}^8 \simeq \mathbb{C}^4$.

Therefore, if Φ is a torsion-free Spin(7)*–structure on an* 8*–manifold, then* $\mathrm{Hol}^0(g_\Phi)$ *is one of* {1}, SU(2), SU(2) × SU(2), SU(3), G_2, Sp(2), SU(4) *or* Spin(7).

We give two particularly interesting instances of the use of these inclusions to obtain Spin(7)–manifolds (with holonomy strictly contained in Spin(7)).

Example 2.41 (Spin(7)–manifolds from G_2–manifolds)**.** Let (Y^7, ϕ) be a G_2–manifold. Let t be a coordinate on $\mathbb{R}$ or S^1. Then the 4–form

$$\Phi := \mathrm{d}t \wedge \phi + \psi,$$

where $\psi = *_Y \phi$, defines a torsion-free Spin(7)–structure on $X^8 := \mathbb{R} \times Y$ or $S^1 \times Y$, compatible with the canonical product metric and orientation.

Example 2.42 (Spin(7)–manifolds from Calabi–Yau 4–folds)**.** Let (Z^8, ω, Υ) be a Calabi–Yau 4–fold. Then the 4–form

$$\Phi := \frac{1}{2}\omega \wedge \omega + \mathrm{Re}(\Upsilon)$$

defines a torsion-free Spin(7)–structure on Z^8 compatible with its metric and orientation.

2.2 Calibrated Geometry

This section is based on Joyce (ibid., §4.1-4.2), Harvey and Lawson (1982), and the lecture notes Lotay (2014) and Nordström (2012).

2.2.1 Minimal Submanifolds

We give a brief recap on the basic definitions concerning minimal submanifolds. We follow the exposition of Joyce's book Joyce (2007, §4.1) and also Lotay's lecture notes Lotay (2014). A classical good reference on this subject is Lawson's lecture notes Lawson (1980).

Definition 2.43 (Submanifold). Let M be smooth manifold. A **submanifold** of M is a one-to-one immersion $\iota\colon N \hookrightarrow M$, where N is some smooth manifold. When N is oriented, we say that $\iota\colon N \hookrightarrow M$ is an *oriented submanifold*. Two submanifolds $\iota\colon N \hookrightarrow M$ and $\iota'\colon N' \hookrightarrow M$ are **isomorphic** if there exists a diffeomorphism $\varphi\colon N \to N'$ such that $\iota = \iota' \circ \varphi$.

We regard isomorphic submanifolds as the same object. In particular, endowing $\iota(N)$ with the manifold structure of N via ι, we do not distinguish between the submanifolds $\iota\colon N \hookrightarrow M$ and $\iota(N) \hookrightarrow M$ (i.e. one can think of N as a subset of M whose inclusion map is ι).

Remark 2.44. We do not require a submanifold $\iota\colon N \hookrightarrow M$ to have the induced topology of the ambient manifold M, i.e. $\iota\colon N \hookrightarrow M$ is not necessarily a topological embedding. Anyway, by the implicit function theorem, we know that any point $p \in N$ has an open neighborhood V such that $\iota|_V$ is a topological embedding. Thus, when addressing local questions, we can suppose N is an embedded submanifold of M. $\Diamond$

In order to formulate the variational approach to minimal submanifolds including noncompact submanifolds, we need the following:

Definition 2.45 (Variations with compact support). Let $\iota\colon N \hookrightarrow M$ be a submanifold and let $S \subseteq N$ be an open subset whose closure in N is compact. A (smooth) **variation** of ι *supported in S* is a smooth map

$$F\colon N \times {]}{-}1, 1{[} \to M$$

such that, writing $\iota_t := F(\cdot, t)$, the following holds:

(i) $\iota_0 = \iota$;

(ii) $\iota_t\colon N \to M$ is a submanifold, for all t;

(iii) $\iota_t|_{N \setminus S} \equiv \iota|_{N \setminus S}$, for all t.

In this case, $V_F \in \mathfrak{X}(N)$ defined by

$$V_F(p) := F(p, \cdot)_* \frac{\partial}{\partial t}\Big|_{t=0}, \quad \forall p \in N,$$

is called the **variational vector field** associated to the variation $F = \{\iota_t\}$.

Definition 2.46 (Minimal submanifolds). Let (M, g) be a Riemannian manifold and let $\iota\colon N \hookrightarrow M$ be an oriented submanifold of M. Denote by $\mathrm{d}V_{\iota_* g}$

the induced Riemannian volume form on N. Then, for each precompact open set $S \Subset N$, it is well-defined the **volume of** S with respect to $\iota\colon N \hookrightarrow M$:

$$\mathrm{Vol}(\iota|_S) := \int_S \mathrm{d}V_{\iota^* g} < \infty.$$

We say that $\iota\colon N \hookrightarrow M$ is a **minimal submanifold** of M when for each precompact open subset [4] $S \Subset N$ we have

$$\frac{\mathrm{d}}{\mathrm{d}t} \mathrm{Vol}(\iota|_S)\Big|_{t=0} = 0,$$

for all variations $\{\iota_t\}$ of ι supported in S.

Thus, a minimal submanifold of M is just a stationary point, with respect to compactly supported variations, of the natural volume functional on oriented submanifolds of M. We can also give a p.d.e. approach to minimal submanifolds by means of the *mean curvature vector* of submanifolds:

Definition 2.47 (Second fundamental form and the mean curvature vector). Let (M, g) be a Riemannian manifold and let $\iota\colon N \hookrightarrow M$ be a submanifold of M. Then, the tangent bundle of M restricted to N decomposes orthogonally as

$$\iota^* TM = \iota_* TN \oplus \nu^\iota(N),$$

where $\nu^\iota(N)$, called the **normal bundle** of $\iota\colon N \hookrightarrow M$, is the vector subbundle of $\iota^* TM$ whose fibre at a point $q \in N$ is the orthogonal complement of $\iota_* T_q N$ in $(\iota^* TM)_q \simeq T_q M$ with respect to g. The **second fundamental form** of $\iota\colon N \hookrightarrow M$ is the section B^ι of $\left(\odot^2 T^* N\right) \otimes \nu^\iota(N)$ such that, for all $X, Y \in \mathfrak{X}(N)$,

$$B^\iota(X, Y) = \pi_{\nu^\iota(N)} \circ \left[D^g_{\iota_* X}(\iota_* Y) \right],$$

where $\pi_{\nu^\iota(N)}\colon \iota^* TM \to \nu^\iota(N)$ is the orthogonal projection map.

The **mean curvature vector** H^ι of $\iota\colon N \hookrightarrow M$ is the section of $\nu^\iota(N)$ given by

$$H^\iota := \mathrm{tr}_{\iota^* g} B^\iota.$$

A straightforward calculation gives the following characterisation of minimal submanifolds Lawson (1980, Theorem 1).

Theorem 2.48. $\iota\colon N \hookrightarrow M$ *is a minimal submanifold of* (M, g) *if, and only if,* $H^\iota \equiv 0$.

---

[4] If $\partial N \neq \emptyset$ one requires S to be in the interior of N.

Note that, by the definition, B^ι depends nonlinearly on the second derivatives of ι, thus so does H^ι. Therefore, the above theorem implies the minimal submanifold condition can be seen as a (nonlinear) p.d.e. of second order on ι, namely, $H^\iota \equiv 0$.

Example 2.49. For immersed curves $\gamma \colon I \to M$, the zero mean curvature condition $H^\gamma \equiv 0$ is equivalent to the geodesic equation $(\gamma^* D^g)(\dot\gamma) = 0$.

Example 2.50. Let $f \colon U \subseteq \mathbb{R}^k \to \mathbb{R}^{n-k}$ be a smooth map from an open subset U of $\mathbb{R}^k$. Then, the graph $\Gamma(f)$ of f is a submanifold of $\mathbb{R}^n$ by means of the natural inclusion map $\iota \colon \Gamma(f) \hookrightarrow \mathbb{R}^k \times \mathbb{R}^{n-k}$. One can show that $H^\iota = 0$ (i.e. $\iota \colon \Gamma(f) \hookrightarrow \mathbb{R}^n$ is a minimal submanifold) if, and only if,

$$\mathrm{div}\left(\frac{\mathrm{grad}(f)}{\sqrt{1 + |\mathrm{grad}(f)|^2}} \right) = 0.$$

2.2.2   Calibrated Submanifolds

The notion of calibration was introduced by Harvey–Lawson (1982) in their seminal paper Harvey and Lawson (1982). Throughout this section, (M^n, g) will be a Riemannian manifold. For each $x \in M$, we denote by $\mathrm{Gr}_+(k, T_x M)$ the **Grassmannian of oriented k−planes** in $T_x M$, i.e.

$$\mathrm{Gr}_+(k, T_x M) := \{V \leqslant T_x M : V \text{ is an oriented } k\text{−subspace of } T_x M\},$$

and we set

$$\mathrm{Gr}_+(k, TM) := \bigcup_{x \in M} \mathrm{Gr}_+(k, T_x M).$$

Elements of $\mathrm{Gr}_+(k, TM)$ are called **oriented tangent k-planes** of M. Note that g induces an inner product $g|_V$ on each $V \in \mathrm{Gr}_+(k, TM)$, which, together with the orientation of V, gives rise to a preferred volume form vol_V on V. In particular, for each $x \in M$, we get an inclusion

$$\mathrm{Gr}_+(k, T_x M) \hookrightarrow \Lambda^k T_x M$$

mapping each $V \in \mathrm{Gr}_+(k, T_x M)$ into the unit simple k−vector $\xi_V := e_1 \wedge \ldots \wedge e_k$, where $\{e_i\}$ is any oriented orthonormal basis of V.

   Recall that each $\phi_x \in \Lambda^k T_x^* M$ defines a linear functional $\langle \phi_x, \cdot \rangle \colon \Lambda^k T_x M \to \mathbb{R}$ by means of the natural pairing

$$\langle \cdot, \cdot \rangle \colon \Lambda^k T_x^* M \otimes \Lambda^k T_x M \to \mathbb{R},$$

defined on simple elements by

$$\langle \alpha_1 \wedge \ldots \wedge \alpha_k, v_1 \wedge \ldots \wedge v_k \rangle := \det\left(\alpha_i(v_j)\right).$$

For $\phi \in \Omega^k(M)$, the **comass** of ϕ at $x \in M$ is the quantity

$$\|\phi\|_x^* := \sup\{\langle \phi_x, \xi_V \rangle : V \in \mathrm{Gr}_+(k, T_x M)\}.$$

More generally, if $A \subseteq M$ is any subset, we define the comass of ϕ on A by

$$\|\phi\|_A^* := \sup\{\|\phi\|_x^* : x \in A\}.$$

When $A = M$, we simply write $\|\phi\|^*$ for the comass of ϕ on M.

When $\phi \in \Omega^k(M)$ and $V \in \mathrm{Gr}_+(k, TM)$, the restriction $\phi|_V$ is a scalar multiple $\lambda_V \in \mathbb{R}$ of vol_V by dimension reasons. In case $\lambda_V \leqslant 1$, we write $\phi|_V \leqslant \mathrm{vol}_V$. Note that

$$\|\phi\|_x^* = \sup\{\lambda_V : V \in \mathrm{Gr}_+(k, T_x M)\}.$$

In particular, $\|\phi\|^* \leqslant 1$ if, and only if, $\phi|_V \leqslant \mathrm{vol}_V$ for all $V \in \mathrm{Gr}_+(k, TM)$.

Definition 2.51. A $k-$form $\phi \in \Omega^k(M)$ is called a **calibration** on (M, g) if

(i) (ϕ is closed) $\mathrm{d}\phi = 0$.

(ii) (ϕ has comass $\leqslant 1$) $\|\phi\|^* \leqslant 1$.

In this case, we define the $\phi-$**Grassmannian** $\mathscr{G}(\phi)$ as the collection of oriented tangent $k-$planes of M where ϕ assumes its maximum, i.e.

$$\mathscr{G}(\phi) := \{V \in \mathrm{Gr}_+(k, TM) : \phi|_V = \mathrm{vol}_V\}.$$

An element $V \in \mathscr{G}(\phi)$ is called a $\phi-$**calibrated** (tangent) $k-$plane.

Remark 2.52. Under the Euclidean identification $\Lambda^k \mathbb{R}^n \simeq \Lambda^k(\mathbb{R}^n)^*$, embed $\mathrm{Gr}_+(k, \mathbb{R}^n) \hookrightarrow \Lambda^k(\mathbb{R}^n)^*$. The Hodge star operator gives isometries $*: \Lambda^k(\mathbb{R}^n)^* \to \Lambda^{n-k}(\mathbb{R}^n)^*$ and $*: \Lambda^k \mathbb{R}^n \to \Lambda^{n-k} \mathbb{R}^n$. Then, for an oriented $k-$plane $V \in \mathrm{Gr}_+(k, \mathbb{R}^n)$ its Hodge star dual $*V$ is the unique orthogonal oriented $(n-k)-$plane $V^\perp$ such that if $\phi \in \Lambda^k(\mathbb{R}^n)^*$ with $\phi|_V = \alpha \mathrm{vol}_V$ for $\alpha \in \mathbb{R}$ then $*\phi|_{V^\perp} = \alpha \mathrm{vol}_{V^\perp}$.

This has the following consequence. Fixing an orientation on (M, g), let $\phi \in \Omega^k(M)$ be a harmonic form (i.e. $\mathrm{d}\phi = 0 = \mathrm{d}*\phi$). Then, ϕ is a calibration if, and only if, $*\phi$ is a calibration, and in this case we have further $*\mathscr{G}(\phi) = \mathscr{G}(*\phi)$. $\Diamond$

Example 2.53. Any $k-$form $\phi \neq 0$ on $\mathbb{R}^n$ with constant coefficients (hence $d\phi = 0$) can be rescaled so that it becomes a calibration with at least one oriented $k-$plane $V_0 \leqslant \mathbb{R}^n$ for which $V_0 \in \mathscr{G}(\phi)$. Indeed, since $\mathrm{Gr}_+(k, \mathbb{R}^n)$ is compact, the comass $\kappa := \|\phi\|_{\mathbb{R}^n}^* \neq 0$ of ϕ on $\mathbb{R}^n$ is attained at some oriented $k-$plane $V_0 \leqslant \mathbb{R}^n$. Thus, whenever $\lambda' \leqslant \lambda := 1/\kappa$, the $k-$form $\lambda'\phi$ is a calibration in $\mathbb{R}^n$, and in case $\lambda' = \lambda$ we have $\lambda\phi|_{V_0} = \mathrm{vol}_{V_0}$.

Although there are usually many calibrations, as the above example shows, it may occur that a calibration just admits a few calibrated tangent $k-$planes, i.e. the $\phi-$Grassmannian $\mathscr{G}(\phi)$ may be 'too small'. The interesting calibrations are the ones for which $\mathscr{G}(\phi)$ is 'big enough' to distinguish a meaningful collection of $k-$submanifolds of M whose tangent spaces lie in $\mathscr{G}(\phi)$. This motivates the following.

Definition 2.54. Let $\phi \in \Omega^k(M)$ be a calibration on (M, g). If $\iota \colon N \hookrightarrow M$ is an oriented $k-$dimensional submanifold of M, then N is called a $\phi-$**calibrated submanifold** (or a $\phi-$**submanifold** for short) when $\iota_* TN \subseteq \mathscr{G}(\phi)$ (as bundles), i.e. when

$$\iota^* \phi = dV_{\iota^* g},$$

where $dV_{\iota^* g}$ is the Riemannian volume form on N induced by $\iota^* g$ and the orientation of N. The collection of $\phi-$submanifolds of M is called the $\phi-$**geometry** of M.

Asking for $\phi-$submanifolds greatly restricts the calibrations ϕ one wants to consider. The next result gives us a key distinguished property of $\phi-$submanifolds (at least in the compact case).

Proposition 2.55. *Let $\phi \in \Omega^k(M)$ be a calibration on (M, g). If $\iota \colon N \hookrightarrow M$ is a compact $\phi-$submanifold, then its volume is the topological invariant $\langle [\iota^* \phi], [N] \rangle$, and it is a minimal submanifold, minimizing volume in its homology class.*

Proof. Let $\iota' \colon N' \hookrightarrow M$ be another compact oriented $k-$submanifold of M such that $\partial N = \partial N'$ and $[N] = [N']$ in $H_k(M, \mathbb{R})$ (i.e. $N - N' = \partial X$, for some $(k+1)-$submanifold X of M). Then,

$$\mathrm{Vol}(\iota) := \int_N dV_{\iota^* g} = \int_N \iota^* \phi = \int_{N'} (\iota')^* \phi \leqslant \int_{N'} dV_{(\iota')^* g} =: \mathrm{Vol}(\iota'),$$

where the second equality follows from the condition of N being a $\phi-$submanifold, in the third equality we used the homology condition on N' together with Stokes' theorem and the fact that ϕ is closed, and in the last inequality we used the fact that ϕ has comass $\leqslant 1$.

To see that this implies $\iota\colon N \hookrightarrow M$ is a minimal submanifold, note that for small t a variation $\iota_t\colon N \hookrightarrow M$ (cf. Definition 2.45) of ι determines the same homology class inside M. Thus, the last inequality shows that

$$\mathrm{Vol}(\iota) \leqslant \mathrm{Vol}(\iota_t),$$

so that ι is a critical point of the volume functional on compact oriented $k-$submanifolds. $\qquad\square$

Corollary 2.56. *There are no compact calibrated submanifolds in a contractible Riemannian manifold (M, g). (e.g. $(\mathbb{R}^n, g_0)$)*

Proof. Let $1 \leqslant k \leqslant n$. By Poincaré's lemma, if $\phi \in \Omega^k(M)$ is a calibration, then there exists $\eta \in \Omega^{k-1}(M)$ such that $\phi = \mathrm{d}\eta$ (indeed, $\mathrm{d}\phi = 0$). Thus, if $\iota\colon N \hookrightarrow M$ is a compact (without boundary) $\phi-$submanifold, using Stokes' theorem we get a contradiction:

$$0 < \mathrm{Vol}(\iota) = \int_N \iota^*\phi = \int_N \mathrm{d}(\iota^*\eta) = 0. \qquad\square$$

The $\phi-$submanifold condition (Definition 2.54) for an oriented compact $k-$dimensional submanifold $\iota\colon N \hookrightarrow M$ depends upon its tangent spaces; it is a *first* order p.d.e. on the immersion ι. On the other hand, as we have already seen in the previous section (Theorem 2.48), the minimal submanifold condition for such a submanifold turns out to be a *second* order p.d.e. on the immersion ι ($H^\iota \equiv 0$). This suggests, via Proposition 2.55, that calibrated geometry is a great source of examples of minimal submanifolds. This fact is quite analogous to the relation, in the realm of gauge theory, between ASD instantons and Yang–Mills connections (Section 1.5). Indeed, in the next section we shall extend this analogy, by means of the general notion of $\Xi-$ASD instantons.

Furthermore, we shall see a striking *concrete* relation between gauge theory and calibrated geometries in dimensions greater than four (cf. Tian (2000)). This will require generalising the notion of $\phi-$submanifold in M to the more general measure-geometric setting of currents on M. In what follows we use some notation and terminology which are introduced in Appendix A (see §A.6).

Definition 2.57 ($\phi-$currents). Let $\phi \in \Omega^k(M)$ be a calibration on (M, g). Then an integral $k-$current $T = (\Gamma, \xi, \Theta) \in \mathbf{I}_k(M)$ (cf. Definition A.67) is said to be a $\phi-$**calibrated current** (or simply $\phi-$**current**) if

$$\phi|_{T_x\Gamma} = \xi(x), \quad \text{for } \mathcal{H}^k - \text{a.e. } x \in \Gamma.$$

Definition 2.58 (Mass-minimizing currents). A current $T \in \mathbf{I}_{k,\mathrm{loc}}(M)$ is called **mass-minimizing** if

$$\mathbf{M}(S) \leqslant \mathbf{M}(S')$$

whenever $S, S' \in \mathbf{I}_k(M)$, $\|T\| = \|S\| + \|T - S\|$ (i.e. S is a *piece* of T) and $\partial S = \partial S'$.

We have the following result in parallel with Proposition 2.55.

Proposition 2.59. *Let $\phi \in \Omega^k(M)$ be a calibration on (M, g). Then any compactly supported $\phi-$calibrated cycle $T \in \mathscr{Z}_k(M) \subseteq \mathbf{I}_k(M)$ is mass-minimizing in its homology class.*

Proof. Write $T = (\Gamma, \xi, \Theta)$ and let $T' = (\Gamma', \xi', \Theta') \in \mathscr{Z}_k(M)$ be a compactly supported cycle homologous to T, say $T - T' = \partial R$, where $R \in \mathbf{I}_{k+1}(M)$. Then, unraveling definitions, we have:

$$\mathbf{M}(T) = \int_{\Gamma} \Theta \mathrm{d}\mathcal{H}^k = \int_{\Gamma} \langle \phi, \xi \rangle \Theta \mathrm{d}\mathcal{H}^k \quad (T \text{ is } \phi-\text{calibrated})$$

$$= \int_{\Gamma'} \langle \phi, \xi' \rangle \Theta' \mathrm{d}\mathcal{H}^k + R(\mathrm{d}\phi) \quad (T - T' = \partial R)$$

$$\leqslant \int_{\Gamma'} \Theta' \mathrm{d}\mathcal{H}^k = M(T'). \quad (\|\phi\|^* \leqslant 1) \qquad \square$$

In this context, it is worth mentioning the following deep interior regularity result due to Almgren (1984).

Theorem 2.60 (Almgren). *If $T \in \mathbf{I}_{k,loc}(M)$ is mass-minimizing, then $\mathring{T} := \mathrm{supp}(T) \setminus \mathrm{supp}(\partial T)$ is a smooth $k-$dimensional minimal submanifold of M, except by a singular set $\Sigma \subseteq \mathring{T}$ of Hausdorff dimension at most $k - 2$.*

Calibrations and Riemannian holonomy groups.   There is a natural method to construct interesting calibrations ϕ on Riemannian manifolds (M, g) with special holonomy, in such a way that $\mathscr{G}(\phi)$ contains families of calibrated tangent $k-$planes with reasonably large dimension.

Let $H \subseteq \mathrm{SO}(n)$ be a possible holonomy group for a Riemannian metric. Thus H acts on the $k-$forms $\Lambda^k(\mathbb{R}^n)^*$ of $\mathbb{R}^n$. Suppose that $\phi_0 \in \Lambda^k(\mathbb{R}^n)^*$ is a nonzero $H-$invariant $k-$form on $\mathbb{R}^n$. Up to rescaling, we can assume that $\|\phi\|^* \leqslant 1$ and that $\mathscr{G}(\phi_0) \neq \emptyset$, i.e. $\phi_0|_V = \mathrm{vol}_V$ for at least one $k-$plane $V \leqslant \mathbb{R}^n$ (see Example 2.53). Thus, from the $H-$invariance of ϕ_0, if $V \in \mathscr{G}(\phi_0)$ then $h \cdot V \in \mathscr{G}(\phi_0)$ for *every* $h \in H$. This usually means $\mathscr{G}(\phi_0)$ is reasonably big.

Now suppose (M, g) is a connected Riemannian $n-$manifold with $\mathrm{Hol}(g) = H$. Then, by the holonomy principle (Theorem 1.32), there exists a global parallel (hence *closed*) $k-$form ϕ on M which is pointwise linearly identified with ϕ_0. It follows that ϕ also has comass $\leqslant 1$ and, therefore, is a *calibration* on M. Moreover, for each $x \in M$, we have $\mathscr{G}(\phi) \cap T_x M \simeq \mathscr{G}(\phi_0)$, so that by the above invariance we may expect the $\phi-$geometry of M is non-trivial.

In what follows, we explore the above procedure for the holonomy groups $\mathrm{U}(m)$, $\mathrm{SU}(m)$, G_2 and $\mathrm{Spin}(7)$, introducing corresponding interesting calibrated geometries.

Complex submanifolds. Let $H = \mathrm{U}(m) \subseteq \mathrm{SO}(2m)$. Then H preserves the standard Kähler 2−form ω_0 on $\mathbb{R}^{2m}$. The following classical lemma shows that $\omega_0^k / k!$ has comass $\leqslant 1$ for each $1 \leqslant k \leqslant m$ Lawson (1980, Proposition 4).

Lemma 2.61 (Wirtinger's inequality). *Consider* $\mathbb{C}^m = \mathbb{R}^{2m}$ *with complex coordinates* $z^j = x^{2j-1} + i x^{2j}$, $j = 1, \dots, m$, *and let* ω_0 *be the standard Kähler form*

$$\omega_0 = \frac{i}{2} \sum_{j=1}^{m} \mathrm{d}z^j \wedge \mathrm{d}\overline{z^j} = \sum_{j=1}^{m} \mathrm{d}x^{2j-1} \wedge \mathrm{d}x^{2j}.$$

Then, for each $1 \leqslant k \leqslant n$, *given any collection of* $2k$ *unitary vectors* $v_1, \dots, v_{2k} \in \mathbb{R}^{2m}$, *we have*

$$\frac{\omega_0}{k!}(v_1, \dots, v_{2k}) \leqslant 1.$$

Corollary 2.62. *Let* (Z^{2m}, I, ω) *be a Kähler* $m-$*fold. Then, for each* $1 \leqslant k \leqslant m$, *the* $2k-$*form* $\dfrac{\omega^k}{k!}$ *is a calibration on* Z. *Moreover, an oriented real* $2k-$*submanifold* N *in* Z *is calibrated if, and only if,* N *is a complex* $k-$*dimensional submanifold of* (Z^{2m}, I), *i.e.* $I(T_x N) = T_x N$ *for all* $x \in N$.

There are lots of examples in this setting. For instance, the complex projective spaces $\mathbb{CP}^m$ have many complex submanifolds defined as the zero set of a collection of homogeneous polynomials. These are called *complex algebraic varieties*, and are the subject of *complex algebraic geometry*. It is worth mentioning that the motivation for the general calibration condition comes from the long-known properties enjoyed by complex submanifolds as minimal submanifolds of Kähler manifolds. For more details and examples of complex submanifolds in Kähler manifolds, we refer the reader to Lawson (ibid., Chapter 1, §6).

Special Lagrangians.   Let $H = \mathrm{SU}(m) \subseteq \mathrm{SO}(2m)$. Then H preserves not only the standard Kähler form ω_0 but also the holomorphic volume form Υ_0. It turns out that $\mathrm{Re}(\Upsilon_0)$ is a calibration on $\mathbb{C}^m$. In fact, the following holds Harvey and Lawson (1982, Theorem 1.14, p. 89):

Lemma 2.63. *Consider* $\mathbb{C}^m = \mathbb{R}^{2m}$ *with complex coordinates* $(z^1, \ldots, z^m)$, *let* ω_0 *be the standard Kähler form and let* Υ_0 *be the holomorphic volume form*

$$\Upsilon_0 := \mathrm{d}z^1 \wedge \ldots \wedge \mathrm{d}z^m.$$

Then

$$|\Upsilon(e_1, \ldots, e_m)| \leqslant 1,$$

for all unit vectors $e_1, \ldots, e_m \in \mathbb{C}^m$ *with equality if, and only if,* $V = \mathrm{span}_{\mathbb{R}}\{e_1, \ldots, e_n\}$ *is a* Lagrangian *plane, i.e.* $\omega_0|_V \equiv 0$.

Corollary 2.64. *Let* $(Z^{2m}, \omega, \Upsilon)$ *be a Calabi–Yau* $m-$*fold. Then* $\mathrm{Re}(e^{\mathrm{i}\theta}\Upsilon)$ *is a calibration on* Z *for any* $\theta \in \mathbb{R}$.

Definition 2.65. Let $(Z^{2m}, \omega, \Upsilon)$ be a Calabi–Yau $m-$fold, and let L be an oriented real $m-$submanifold of Z. We call L a **special Lagrangian** submanifold (or SL $m-$fold for short) if L is calibrated with respect to $\mathrm{Re}(\Upsilon)$. More generally, if L is calibrated with respect to $\mathrm{Re}(e^{\mathrm{i}\theta}\Upsilon)$, for some real number $\theta \in \mathbb{R}$, then L is called **special Lagrangian with phase** $e^{\mathrm{i}\theta}$.

Remark 2.66. Let L be an oriented real $m-$submanifold of $\mathbb{C}^m$. Then it is easy to see that L is a SL $m-$fold with phase $e^{\mathrm{i}\theta}$ if, and only if, $e^{-\mathrm{i}\theta}L$ is a SL $m-$fold. $\diamond$

By Lemma 2.63, a $m-$submanifold L of a Calabi–Yau $m-$fold $(Z^{2m}, \omega, \Upsilon)$ admits an orientation making it a SL $m-$fold (with phase 1) if, and only if, $\omega|_L \equiv 0$ (i.e. L is Lagrangian) and $\mathrm{Im}(\Upsilon)|_L \equiv 0$. More generally, from Remark 2.66, it follows that L admits an orientation making it a special Lagrangian with phase $e^{\mathrm{i}\theta}$ if, and only if, $\omega|_L \equiv 0$ and $(\cos\theta\,\mathrm{Im}(\Upsilon) + \sin\theta\,\mathrm{Re}(\Upsilon))|_L \equiv 0$.

For more on special Lagrangian geometry, including several examples, we refer the reader to Joyce (2007, Chapter 8).

Associative and coassociative submanifolds.   Let $H = \mathrm{G}_2 \subseteq \mathrm{SO}(7)$. The next result follows from Harvey and Lawson (1982, Theorem 1.4, p. 113) and Remark 2.52.

Lemma 2.67. *The* $3-$*form* ϕ_0 *given by (2.15) and the* $4-$*form* $\psi_0 = *\phi_0$ *are calibrations on* $\mathbb{R}^7$.

Corollary 2.68. *Let (Y, g_ϕ) be a G_2—manifold. Then ϕ and $\psi = *\phi$ are calibrations.*

Definition 2.69. An oriented 3—submanifold P (resp. 4—submanifold Q) of Y is called **associative** (resp. coassociative) if P (resp. Q) is a ϕ—calibrated (resp. ψ—calibrated) submanifold (cf. Definition 2.54).

Example 2.70. Consider the model G_2—manifold $(\mathbb{R}^7, \phi_0)$ of Example 2.25 and the natural orthogonal decomposition $\mathbb{R}^7 = \mathbb{R}^3 \oplus \mathbb{R}^4$, as in Remark 2.16. Then, from the definition (2.15) of ϕ_0, with the natural choices of orientation - as prescribed in Remark 2.16, it is easy to see that

$$P := \mathbb{R}^3 \times \{0\} \subseteq \mathbb{R}^3 \oplus \mathbb{R}^4 = \mathbb{R}^7 \text{ is associative and}$$
$$Q := \{0\} \times \mathbb{R}^4 \subseteq \mathbb{R}^3 \oplus \mathbb{R}^4 = \mathbb{R}^7 \text{ is coassociative.}$$

More generally, if (Z^4, ω, Υ) is a Calabi–Yau 2—fold and we let $(Y^7 := \mathbb{R}^3 \times Z, \phi)$ be the G_2—manifold of Example 2.30, then (with the obvious orientations)

$$P := \mathbb{R}^3 \times \{0\} \subseteq \mathbb{R}^3 \oplus Z = Y \text{ is associative and}$$
$$Q := \{0\} \times Z \subseteq \mathbb{R}^3 \oplus Z = Y \text{ is coassociative.}$$

The following result, which we state without proof, gives us a good source of examples of associative and coassociative submanifolds.

Proposition 2.71. *Let (Y, ϕ) be a G_2—manifold with an isometric involution $\sigma \neq \mathbb{1}$. If $\sigma^*\phi = \phi$ (resp. if $\sigma^*\phi = -\phi$), then*

$$\mathrm{Fix}(\phi) := \{p \in M : \sigma(p) = p\}$$

is a closed embedded associative (resp. coassociative) submanifold in Y.

We refer the reader to Joyce (2007, pp. 268-269) for a proof of the above result, as well as examples of associative and coassociative submanifolds arising in this way.

Example 2.72. The recent work by Corti et al. (2015) contains various concrete examples of associative submanifolds in G_2—manifolds, arising from the twisted connected sum construction.

Next we state a reduction result to lower-dimensional calibrated geometries. Its proof follows rather easily from the compatibility of the involved structures.

Proposition 2.73. *Let (Z^6, ω, Υ) be a Calabi–Yau 3–fold and consider the cylindrical G_2–manifold $(Y := \mathbb{R} \times Z, \phi)$ of Example 2.31. Then:*

(a) *$N := \mathbb{R} \times \Sigma \subseteq Y$ associative (resp. coassociative) if and only if Σ is a complex curve (resp. special Lagrangian 3–fold with phase $-i$).*

(b) *$N \subseteq \{x\} \times Z \subseteq Y$ is associative (resp. coassociative) if and only if N is a special Lagrangian 3–fold (resp. complex surface).*

Cayley submanifolds.   Let $H = \mathrm{Spin}(7) \subseteq \mathrm{SO}(8)$.

Lemma 2.74 (Harvey and Lawson (1982, Theorem 1.24))**.** *The 4–form Φ_0 given by (2.33) is a calibration on $\mathbb{R}^8$.*

Corollary 2.75. *Suppose (X^8, Φ) is a $\mathrm{Spin}(7)$–manifold. Then Φ is a calibration on (X, g_Φ).*

Definition 2.76. An oriented 4–submanifold L of X is called **Cayley** if L is a Φ–submanifold.

We can produce examples of Cayley 4–folds from lower dimensional simpler calibrations cf. Examples 2.41 and 2.42.

Example 2.77. Let (Y^7, ϕ) be a G_2–manifold and consider the $\mathrm{Spin}(7)$–manifold $(X^8 := \mathbb{R} \times Y^7, \Phi)$ given by Example 2.41. Then:

(i) L is an associative 3–fold in Y if and only if $\mathbb{R} \times L$ is Cayley in X.

(ii) For each $x \in \mathbb{R}$, L is a coassociative 4–fold in Y if and only if $\{x\} \times L$ is Cayley in X.

Example 2.78. Let (Z^8, ω, Υ) be a Calabi–Yau 4–fold and consider the $\mathrm{Spin}(7)$–manifold $\left(X^8 := Z^8, \Phi := \dfrac{1}{2}\omega \wedge \omega + \mathrm{Re}(\Upsilon_0) \right)$ of Example 2.42. Then:

(i) L is a holomorphic surface in Z if and only if L is Cayley in X.

(ii) L is a special Lagrangian 4–fold in Z if and only if L is Cayley in X.

2.3   Anti-selfduality in higher dimensions

We present two well established approaches to the notion of instanton in higher dimensions, which in fact coincide for (connected) Riemannian manifolds (M^n, g) whose holonomy group $\mathrm{Hol}(g)$ is one of the following: $\mathrm{U}(m)$ $(n = 2m \geqslant 4)$, G_2 $(n = 7)$ and $\mathrm{Spin}(7)$ $(n = 8)$.

Instantons via closed $(n-4)$-forms. This approach was originally explored by physicists in Corrigan et al. (1983), for flat spaces; see also Baulieu, Kanno, and Singer (1998), and further Tian (2000, Section 1.2). Suppose $n \geqslant 4$ and let (M^n, g) be an oriented Riemannian manifold. Given $\Xi \in \Omega^{n-4}(M)$, we define the following $*$-Hodge-type operator acting on 2-forms:

$$*_\Xi : \Lambda^2 T^* M \to \Lambda^2 T^* M$$
$$\omega \mapsto *(\Xi \wedge \omega). \tag{2.79}$$

We note that $*_\Xi$ is trace-free, self-adjoint and satisfies $*_\Xi = 0$ if, and only if, $\Xi = 0$. Of course, for a given G-bundle $E \to M$, there is also a natural extension of $*_\Xi$ to $\mathfrak{g}_E$-valued 2-forms, acting trivially on the $\mathfrak{g}_E$-component. This leads to the following generalisation of the 4-dimensional notion of anti-selfduality.

Definition 2.80 (Ξ-ASD instantons). Let (M, g) be an oriented Riemannian manifold endowed with a closed $(n-4)$-form $\Xi \in \Omega^{n-4}(M)$. Let $E \to M$ be a G-bundle, where G is a compact Lie group.

(i) Suppose that G is a semi-simple Lie group. In this case, a connection $\nabla \in \mathcal{A}(E)$ on E is called a Ξ-**anti-selfdual instanton** (Ξ-ASD instanton) if

$$*(\Xi \wedge F_\nabla) = -F_\nabla. \tag{2.81}$$

(ii) With no semi-simplicity hypothesis on G, it is convenient to relax the above definition as follows (cf. Walpuski (2013b, Remark 1.90, p. 30)). Recalling the decomposition (1.4), a connection $\nabla \in \mathcal{A}(E)$ on E is called a Ξ-**anti-selfdual instanton** (Ξ-ASD instanton) if the $\mathfrak{g}_E^{(0)}$-component of F_∇, instead of F_∇, satisfies (2.81) and the $\mathfrak{z}(\mathfrak{g})$-component of F_∇ is a $\mathfrak{z}(\mathfrak{g})$-valued harmonic 2-form[5].

Remark 2.82. While (ii) indeed generalises (i), Ξ-ASD instantons on a G-bundle E are essentially equivalent to Ξ-ASD instantons on the associated $G/Z(G)$-bundle $E \times_G G/Z(G)$. In other words, we can always reduce to the semi-simple case (i). $\diamond$

Remark 2.83. A case of particular interest encompassed by (ii) is $G = \mathrm{U}(r)$. For later purposes, we introduce some terminology. The quotient group

---

[5]The $\mathfrak{g}_E^{(0)}$-component of F_∇ is simply its trace-free component F_∇^0, and the $\mathfrak{z}(\mathfrak{g})$-component of F_∇ is simply $\frac{1}{r}\mathrm{tr}(F_\nabla) \otimes \mathbb{1}$. Thus, ∇ is a Ξ-ASD instanton if, and only if, $*(\Xi \wedge F_\nabla^0) = -F_\nabla^0$ and $\mathrm{tr}(F_\nabla)$ is a harmonic 2-form.

$U(r)/Z(U(r)) \simeq U(r)/U(1)$, denoted henceforth by $PU(r)$, is called the **projective unitary group** of rank r. We call E a **$PU(r)$–bundle** when E is the associated bundle $\widetilde{E} \times_{U(r)} PU(r)$ of a $U(r)$–bundle $\widetilde{E}$. $\qquad\qquad \Diamond$

In the classical case of an oriented Riemannian 4–manifold (M, g), there is a natural choice of 0-form Ξ, namely $\Xi = *dV_g = 1$, for which $*_\Xi = *$. Of course, the corresponding Ξ–anti-selfduality notion is precisely the familiar one explained in Section 1.5.

For generic Ξ, the algebraic equation (2.81) is an over-determined system and admits no solutions at all (i.e. -1 need not be an eigenvalue of $*_\Xi$; for instance, when $n = 4$, let Ξ be any constant function $\neq 1$ and -1). In any case, similarly to the classical 4–dimensional notion, if $\nabla \in \mathcal{A}(E)$ is a Ξ–ASD instanton, then ∇ is automatically a Yang–Mills connection. Indeed, for the case (i) of Definition 2.80, this is an immediate consequence of $d\Xi = 0$ and the Bianchi identity (1.21):

$$d_\nabla * F_\nabla = -d_\nabla(\Xi \wedge F_\nabla) = -(\underbrace{d\Xi}_{=0} \wedge F_\nabla + (-1)^{n-4}\Xi \wedge \underbrace{d_\nabla F_\nabla}_{=0}) = 0.$$

As for the general case (ii) of Definition 2.80, we have (compare with Tian (2000, Lemma 1.2.1)):

Proposition 2.84. *Let E be a G–bundle, where G is a compact Lie group, and let $\nabla \in \mathcal{A}(E)$. If ∇ is a Ξ–ASD instanton (Definition 2.80 (ii)) then ∇ is a Yang–Mills connection. Moreover, if $G = U(r)$ and M is closed, we have the following a priori L^2–energy bound on ∇:*

$$\|F_\nabla\|_{L^2}^2 - \frac{1}{r}|\mathrm{tr}(F_\nabla)|_{L^2}^2 = 4\pi^2\left\langle\left(2c_2(E) - \frac{r-1}{r}c_1(E)^2\right) \cup [\Xi], [M]\right\rangle.$$

Proof. Since $d(\mathrm{tr}(F_\nabla)) = \mathrm{tr}(d_\nabla F_\nabla)$ and $\nabla\mathbb{1} = 0$, it follows from the Bianchi identity (1.21) and the Leibniz rule that $d_\nabla F_\nabla^0 = 0$. Furthermore,

$$\begin{aligned}
d_\nabla^* F_\nabla &= d_\nabla^*\left(\frac{1}{r}\mathrm{tr}(F_\nabla) \otimes \mathbb{1}\right) \pm *d_\nabla\left(*F_\nabla^0\right) \\
&= \frac{1}{r}\left(d^*\mathrm{tr}(F_\nabla)\right) \otimes \mathbb{1} \mp *d_\nabla\left(\Xi \wedge F_\nabla^0\right) \quad (F_\nabla^0 \text{ is } \Xi\text{–ASD}) \\
&= 0 \mp *\left(d\Xi \wedge F_\nabla^0 + (-1)^{n-4}\Xi \wedge d_\nabla F_\nabla^0\right) \quad (\mathrm{tr}(F_\nabla) \text{ is harmonic}) \\
&= 0. \quad (d\Xi = 0 \text{ and } d_\nabla F_\nabla^0 = 0)
\end{aligned}$$

This proves that ∇ is Yang–Mills.

Suppose $G = U(r)$ and M is a compact manifold without boundary. First, by (1.47) and (1.48), we have

$$4\pi^2 \left(2c_2(E) - \frac{r-1}{r} c_1(E)^2 \right) = \mathrm{tr}(F_\nabla \wedge F_\nabla) - \frac{1}{r} \mathrm{tr}(F_\nabla) \wedge \mathrm{tr}(F_\nabla). \quad (2.85)$$

Note that the decomposition $F_\nabla = F_\nabla^0 + \frac{1}{r}\mathrm{tr}(F_\nabla) \otimes \mathbb{1}$ is L^2−orthogonal; indeed,

$$\begin{aligned}
\langle F_\nabla^0, \mathrm{tr}(F_\nabla) \otimes \mathbb{1} \rangle_{L^2} &= -\int_M \mathrm{tr}\left(F_\nabla^0 \wedge *(\mathrm{tr}(F_\nabla) \otimes \mathbb{1}) \right) \\
&= -\int_M \mathrm{tr}(F_\nabla^0 \wedge *\mathrm{tr}(F_\nabla)) \\
&= -\int_M \mathrm{tr}(F_\nabla^0) \wedge *\mathrm{tr}(F_\nabla) \\
&= 0.
\end{aligned}$$

Thus

$$\|F_\nabla\|_{L^2}^2 = \|F_\nabla^0\|_{L^2}^2 + \|\tfrac{1}{r}\mathrm{tr}(F_\nabla) \otimes \mathbb{1}\|_{L^2}^2 = \|F_\nabla^0\|_{L^2}^2 + \frac{1}{r}|\mathrm{tr}(F_\nabla)|_{L^2}^2. \quad (2.86)$$

Now, F_∇^0 is Ξ−ASD, so

$$\|F_\nabla^0\|_{L^2}^2 = -\int_M \mathrm{tr}(F_\nabla^0 \wedge *F_\nabla^0) = \int_M \mathrm{tr}(F_\nabla^0 \wedge F_\nabla^0 \wedge \Xi).$$

On the other hand,

$$F_\nabla^0 \wedge F_\nabla^0 = F_\nabla \wedge F_\nabla - \frac{2}{r}\mathrm{tr}(F_\nabla) \wedge F_\nabla + \frac{1}{r^2}\mathrm{tr}(F_\nabla) \wedge \mathrm{tr}(F_\nabla) \otimes \mathbb{1}.$$

Therefore,

$$\begin{aligned}
\|F_\nabla^0\|_{L^2}^2 &= \int_M \mathrm{tr}(F_\nabla \wedge F_\nabla \wedge \Xi) - \frac{2}{r}\int_M \mathrm{tr}(F_\nabla) \wedge \mathrm{tr}(F_\nabla) \wedge \Xi \\
&\quad + \frac{\mathrm{tr}(\mathbb{1})}{r^2} \int_M \mathrm{tr}(F_\nabla) \wedge \mathrm{tr}(F_\nabla) \wedge \Xi \\
&= \int_M \mathrm{tr}(F_\nabla \wedge F_\nabla \wedge \Xi) - \frac{1}{r}\int_M \mathrm{tr}(F_\nabla) \wedge \mathrm{tr}(F_\nabla) \wedge \Xi
\end{aligned}$$

Plugging this last equation in (2.86) and comparing with (2.85) gives the desired result. $\qquad \square$

Remark 2.87. Comparing the previous result, for e.g. $G = SU(r)$, with Proposition 2.55 brings forth various similarities between Ξ−ASD instantons and Ξ−calibrated submanifolds: both are first order solutions of second order Euler–Lagrange equations. Furthermore, these solutions in fact minimise their respective defining (energy/volume) functionals, attaining topological (energy/volume) lower bounds. $\diamond$

The following result, albeit straightforward from linear algebra, underlies the relevance of calibrated submanifolds in the study of gauge theory, via the bubbling phenomena in Chapter 4. In fact, it is the reason why the bubbling locus of a sequence of Ξ−ASD instantons is Ξ−calibrated, and why ASD instantons bubbles off transversely (cf. Theorem B).

Proposition 2.88 (ASD instantons bubbles off transversely). *Suppose $n > 4$ and consider $(\mathbb{R}^n, g_0)$ with the standard flat metric g_0. Let $\Xi \in \Omega^{n-4}(\mathbb{R}^n)$ be a calibration and let $\mathbb{R}^n = \mathbb{R}^{n-4} \oplus \mathbb{R}^4$ be an orthogonal decomposition, with associated projection map $\pi \colon \mathbb{R}^n \to \mathbb{R}^4$. Let E be a G−bundle over $\mathbb{R}^4$ where G is a compact Lie group. If $I \in \mathcal{A}(E)$ is a non-flat connection then the following are equivalent:*

(i) $\nabla := \pi^* I$ is a Ξ−ASD instanton.

(ii) There exists an orientation on $\mathbb{R}^{n-4}$ with respect to which it is calibrated by Ξ and I is an ASD instanton on $\mathbb{R}^4$.

Proof. Let us assume, without loss of generality, that G is semi-simple, so that we are in case (i) of Definition 2.80 (see Remark 2.82).

Let $x^1, \ldots, x^n$ be oriented orthonormal coordinates of $\mathbb{R}^n$ such that $x^1, \ldots, x^{n-4}$ are coordinates for $\mathbb{R}^{n-4}$. Set

$$\Phi_{n-4} := \mathrm{d}x^1 \wedge \ldots \wedge \mathrm{d}x^{n-4} \quad \text{and} \quad \Phi_4 := *\Phi_{n-4} = \mathrm{d}x^{n-3} \wedge \ldots \wedge \mathrm{d}x^n,$$

and

$$\Xi = \alpha \Phi_{n-4} + \Xi_0,$$

for some $\alpha \in \mathbb{R}$ and $\Xi_0 \in \Omega^{n-4}(\mathbb{R}^n)$, such that $\Xi_0|_{\mathbb{R}^{n-4}} = 0$. Then, it is clear that

$$*(\Xi \wedge F_\nabla) = \alpha * (\Phi_{n-4} \wedge F_\nabla). \tag{2.89}$$

(i)$\Rightarrow$(ii): Choose (temporarily) Φ_4 to be the orientation of $\mathbb{R}^4$. Then, from (2.89) and the assumption of (i) we get[6]

$$-F_I = \alpha *_{\mathbb{R}^4} F_I.$$

---

[6]Recall that $F_\nabla = \pi^* F_I$ and that π is a submersion.

Since $F_I \neq 0$, it follows that $\alpha = \pm 1$ (the possible eigenvalues of $*$ on $\Lambda^2(\mathbb{R}^4)^*$). If $\alpha = 1$, we are done. If $\alpha = -1$, just choose the reverse orientation on $\mathbb{R}^4$ (note that this changes $*_{\mathbb{R}^4}$ by a minus sign).

(ii)$\Rightarrow$(i): We can assume that Φ_{n-4} is positively oriented, with respect to the orientation on $\mathbb{R}^{n-4}$ predicted by (ii). For, otherwise, recalling that $n > 4$, we can simply make the coordinate change $(x^1, \ldots, x^{n-4}, x^{n-3}, \ldots, x^n) \mapsto (x^1, \ldots, -x^{n-4}, -x^{n-3}, \ldots, x^n)$.

Thus, by assumption, $\alpha = 1$. Also, we fix the compatible orientation given by Φ_4 on $\mathbb{R}^4$. Then, using (2.89) and the hypothesis that F_I is ASD, we get

$$*(\Xi \wedge F_\nabla) = \pi^*(*_{\mathbb{R}^4} F_I) = -\pi^* F_I = -F_\nabla,$$

as claimed. $\qquad\square$

Appropriate $(n-4)$–forms and Riemannian holonomy groups.
(cf. Alekseevsky, Cortés, and Devchand (2003, pp. 8-9) and S. Salamon (1989, p. 61))

We will show that manifolds with reduced holonomy are an appropriate setting to find natural closed $(n-4)$–forms Ξ, for which the Ξ–ASD criterion (2.81) is meaningful.

On M^n, $n \geq 4$, an **appropriate** $(n-4)$–form $\Xi \in \Omega^{n-4}(M)$ is such that the symmetric operator $*_\Xi$ (2.79) admits -1 as one of its (necessarily real) eigenfunctions.

Suppose $H \subseteq SO(n)$ is a Lie subgroup preserving a nonzero 4–form $\Psi_0 \in \Lambda^4(\mathbb{R}^n)^*$. Then, if M is endowed with a H–structure $\mathcal{P} \subseteq \mathcal{F}(M)$, we automatically get a corresponding well-defined nowhere zero 4–form Φ on M, pointwise linearly identified with Ψ_0. Explicitly, $\Phi \in \Omega^4(M)$ is defined by

$$\Psi_x := (u_x^{-1})^* \Psi_0,$$

for any chosen frame $u_x \in \mathcal{P}_x$, for all $x \in M$. This is well-defined, since any two frames $u_x, \tilde{u}_x \in \mathcal{P}_x$ are related by the right multiplication of an element in H: $\tilde{u}_x = h^{-1} \circ u_x$, for some $h \in H$. Thus, the H–invariance of Ψ_0 ensures $(u_x^{-1})^* \Psi_0 = (\tilde{u}_x^{-1})^* \Psi_0$.

By the same reasoning, since $H \subseteq SO(n)$, it follows that M has the structure of an oriented Riemannian manifold. In particular, we are able to define $\Xi := *\Psi \in \Omega^{n-4}(M)$. Notice that the matrix of $*_\Xi$ with respect to any H–frame $u \in \mathcal{P}$ is constant and equal to the matrix of the operator $*_{\Xi_0}$ acting on $\Lambda^2(\mathbb{R}^n)^*$, where $\Xi_0 := *\Psi_0 \in \Lambda^{n-4}(\mathbb{R}^n)^*$. Since $*_{\Xi_0}$ is a nonzero symmetric operator (indeed, $\Xi_0 \neq 0$), it admits a nonzero eigenvalue $0 \neq \lambda \in \mathbb{R}$. Thus, it follows that $-\lambda^{-1}\Xi$ is an appropriate $(n-4)$–form on M.

There are several examples of subgroups $H \subseteq \mathrm{SO}(n)$ admitting nonzero H−invariant 4−forms. Suppose $H \subseteq \mathrm{SO}(n)$ is a closed Lie subgroup with Lie algebra $\mathfrak{h} \subseteq \mathfrak{so}(n) \simeq \Lambda^2(\mathbb{R}^n)^*$. Then, the Killing form $K_{\mathfrak{h}}$ of $\mathfrak{h}$ can be seen as an H−invariant element of $S^2(\mathfrak{h}) \subseteq S^2\left(\Lambda^2(\mathbb{R}^n)^*\right)$. Thus, we can define a corresponding H−invariant 4−form Ψ_0^H on $\mathbb{R}^n$ by

$$\Psi_0^H := \mathrm{alt}(K_{\mathfrak{h}}) \in \left(\Lambda^4(\mathbb{R}^n)^*\right)^H ,$$

where $\mathrm{alt}\colon S^2\left(\Lambda^2(\mathbb{R}^n)^*\right) \to \Lambda^4(\mathbb{R}^n)^*$ denotes the alternation map. If a manifold M is endowed with a H−structure, we denote by Ψ^H the induced 4−form on M. When H is the holonomy group of a Riemannian manifold, this yields the following result S. Salamon (1989, Lemma 5.3).

Lemma 2.90. *Let (M, g) be a Riemannian manifold with holonomy group $H \subset \mathrm{SO}(n)$. Then, the above procedure defines a nowhere zero parallel 4−form Ψ^H on M, except possibly when H is the isotropy representation of a symmetric space.*

Ψ^H is often called the *fundamental 4−form* associated to the holonomy reduction. Since Ψ^H is D^g−parallel, it follows that it is a harmonic 4−form, so that $\Xi^H := *\Psi^H$ defines a closed $(n-4)$−form on M. As observed earlier, modulo rescaling, Ξ^H defines an appropriate $(n-4)$−form on M.

Now suppose $H \subseteq \mathrm{SO}(n)$ is a simple Lie group, e.g. $H = G_2 \subseteq \mathrm{SO}(7)$ or $\mathrm{Spin}(7) \subseteq \mathrm{SO}(8)$ (cf. Theorems 2.18 and 2.34). Since Ξ_0^H is by construction H−invariant, the operator $*_{\Xi_0^H}$ trivially commutes with the action of H, so by Schur's lemma the irreducible representations of H in $\Lambda^2(\mathbb{R}^n)^*$ are eigenspaces for $*_{\Xi_0^H}$. Since H is simple, it follows that the Lie algebra $\mathfrak{h} \subseteq \mathfrak{so}(n) \simeq \Lambda^2(\mathbb{R}^n)^*$ is an eigenspace for $*_{\Xi_0^H}$.

In the situation of Lemma 2.90, it follows that the natural subbundle $\widetilde{\mathfrak{h}} \subseteq \Lambda^2 T^*M$ determined by $\mathfrak{h}$ is one of the eigenbundles of the operator $*_{\Xi^H}$. If $0 \neq \lambda = \mathrm{const.}$ is the corresponding eigenvalue, then scaling Ξ^H by $-\lambda^{-1}$ yields an appropriate closed $(n-4)$−form $\widetilde{\Xi}^H$ on M, whose -1 eigenbundle is precisely $\widetilde{\mathfrak{h}}$. The corresponding $\widetilde{\Xi}^H$−anti-selfduality notion is then a natural constraint coming from the holonomy reduction of (M, g), in analogy with the constraints imposed by the holonomy group on the Riemann curvature tensor (cf. Proposition 2.6). This leads us to another the notion of instanton in higher dimensions, which we will now briefly describe.

Instantons via Lie groups. There is a generalised notion of instanton available for any oriented Riemannian manifold (M^n, g) equipped with an

$N(H)$−structure Carrión (1998). Let $H \subseteq \mathrm{SO}(n)$ be a closed Lie subgroup. Then we can write

$$\Lambda^2(\mathbb{R}^n)^* \simeq \mathfrak{so}(n) = \mathfrak{h} \oplus \mathfrak{h}^\perp. \tag{2.91}$$

Recall that

$$\Lambda^2 T^* M = \mathcal{F}(M) \times_{\mathrm{SO}(n)} \Lambda^2(\mathbb{R}^n)^*.$$

If M has an H−structure, then the decomposition (2.91) readily goes over to 2−forms on M. However, in practice, one often has a $N(H)$−structure instead of a H−structure, where $N(H) \supseteq H$ denotes the normaliser of H in $\mathrm{SO}(n)$. So, suppose (M, g) has an $N(H)$−structure $\mathcal{P} \subseteq \mathcal{F}(M)$. Since H is closed, it follows that $N(H)$ is a closed Lie subgroup of $\mathrm{SO}(n)$. Moreover, one can easily verify that $\mathfrak{h} \subseteq \mathfrak{n}(\mathfrak{h}) =: \mathrm{Lie}(N(H))$ is invariant under the adjoint action of $N(H)$. This means that the lie algebra $\mathfrak{h}$ determines a distinguished subbundle $\widetilde{\mathfrak{h}}$ of $\Lambda^2 T^* M$:

$$\widetilde{\mathfrak{h}} := \mathcal{P} \times_{N(H)} \mathfrak{h} \subseteq \Lambda^2 T^* M.$$

In particular, if $E \to M$ is a G−bundle, using the metric g we get an associated orthogonal projection map

$$\pi_{\mathfrak{h}} : \Lambda^2 T^* M \otimes \mathfrak{g}_E \to \widetilde{\mathfrak{h}} \otimes \mathfrak{g}_E.$$

Definition 2.92. Suppose M is endowed with a $N(H)$−structure, and let $E \to M$ be a G−bundle with compact semi-simple structure group. A connection $\nabla \in \mathcal{A}(E)$ is called an H−**instanton** if

$$\pi_{\mathfrak{h}} F_\nabla = F_\nabla,$$

i.e. if the curvature 2−forms F^i_j lies in the subspace $\mathfrak{h} \subseteq \Lambda^2$.

For instance, in the classical case of a 4−manifold, M comes equipped with a $N(H) = \mathrm{SO}(4)$−structure, for $H = \mathrm{SU}(2)$. Then a $\mathrm{SU}(2)$−instanton corresponds to the ordinary notion of (A)SD instanton: $\mathfrak{h} = \mathfrak{su}(2) \simeq \Lambda^2_\pm$.

When looking for manifolds endowed with an $N(H)$−structure, a natural guess consists of Riemannian manifolds with reduced holonomy. In particular, we are led to consider each $N(H)$ arising in Berger's list (2.7) of special geometries.

Let (M, g) be an oriented Riemannian n−manifold with $\mathrm{Hol}(g) = N(H) \subsetneq \mathrm{SO}(n)$, where $N(H) \neq \mathrm{SO}(4)$ is one of the groups in Table 2.1 (cf. Carrión (ibid., p. 6, Table 1)). By the construction of the last paragraph, we have a naturally associated parallel 4−form $\Phi^{N(H)}$ arising from the holonomy reduction. It turns out that the corresponding notions of H−instanton

n	H	$N(H) \subseteq SO(n)$
4	SU(2)	SO(4)
$2m > 4$	SU(m)	U(m)
7	G_2	G_2
8	Spin(7)	Spin(7)

Table 2.1: Certain Lie groups $H \subseteq SO(n)$ whose normaliser $N(H)$ in $SO(n)$ is a Lie group appearing in Berger's list.

and $\widetilde{\Xi}^{N(H)}$−ASD instanton induced on auxiliary G−bundles $E \to M$ are co-incident, provided $\widetilde{\Xi}^{N(H)}$ is an appropriate rescaling of $\Xi^{N(H)}$. In the next sections we briefly study each of these cases.

2.3.1   Hermitian–Yang–Mills connections

The following definition is motivated by the discussion in Section 1.5:

Definition 2.93. Let (Z, ω) be a Kähler manifold and let $E \to Z$ be a SU(r)− or a PU(r)−bundle. A connection $\nabla \in \mathcal{A}(E)$ is called **Hermitian–Yang–Mills** (HYM) if

$$F_\nabla^{0,2} = 0 \quad \text{and} \quad \Lambda_\omega F_\nabla = 0. \tag{2.94}$$

Here Λ_ω is the dual of the Lefschetz operator $L_\omega := \omega \wedge \cdot$.

Remark 2.95. Recalling Definition 2.80 (ii) and Remark 2.82, one can also work with U(r)−bundles and, instead of the second part of (2.94), require that $\Lambda_\omega F_\nabla = \lambda \mathbb{1}_E$, for some $\lambda \in i\mathbb{R}$. In this case, if we suppose (Z, ω) is a *compact* (without boundary) Kähler manifold, the value of λ is topologically determined, as follows. Denote by g the underlying Riemannian metric of (Z, ω); thus,

$$\mathrm{d}V_g = \frac{\omega^m}{m!} \quad \text{and} \quad *\,\omega = \frac{\omega^{m-1}}{(m-1)!}.$$

Recalling (1.47), it follows that

$$c_1(E) \wedge *\omega = \lambda r \frac{\omega^m}{m!}.$$

Therefore

$$\lambda = \frac{m \langle c_1(E) \cup [\omega]^{m-1}, [Z] \rangle}{r \langle [\omega], [Z] \rangle}.$$

It follows from Corollary 1.71 that a HYM connection $\nabla \in \mathcal{A}(E)$ induces a holomorphic structure $\mathcal{E}$ on E.

In order to highlight the importance of HYM connections, let us briefly review the so-called Donaldson–Uhlenbeck–Yau correspondence, after Donaldson (1985) and Uhlenbeck and Yau (1986). While a thorough discussion of this result (and its far-reaching repercussions in geometric analysis and algebraic geometry) would be awfully beyond the scope of this text, we kindly refer the interested reader to the excellent books by Kobayashi (2014) and Lübke and Teleman (1995).

Let $\mathcal{E} \to Z$ be a holomorphic vector bundle over the *compact* Kähler manifold (Z, ω). For a coherent subsheaf $\mathcal{F} \subseteq \mathcal{E}$, we define respectively the **first Chern class**, the ω-**degree**, and the ω-**slope**, by [7]

$$c_1(\mathcal{F}) := c_1(\det \mathcal{F}^{**}),$$

$$\deg_\omega(\mathcal{F}) := \int_Z c_1(\mathcal{F}) \wedge \omega^{m-1},$$

$$\mu_\omega(\mathcal{F}) := \frac{\deg_\omega(\mathcal{F})}{\operatorname{rank}(\mathcal{F})}.$$

$\mathcal{E}$ is then called:

- **stable** if $\mu_\omega(\mathcal{F}) < \mu_\omega(\mathcal{E})$, for each coherent subsheaf $\mathcal{F} \subseteq \mathcal{E}$ with $0 < \operatorname{rank}(\mathcal{F}) < \operatorname{rank}(\mathcal{E})$.

- **polystable** if $\mathcal{E} = \bigoplus_i \mathcal{E}_i$ where each $\mathcal{E}_i$ is stable and satisfies $\mu_\omega(\mathcal{E}_i) = \mu_\omega(\mathcal{E})$.

The following deep result was proved by Donaldson (1985) for complex algebraic surfaces, and again by Uhlenbeck and Yau (1986) for compact Kähler manifolds. It gives a very general relation between Yang–Mills theory over Kähler manifolds and Mumford–Takemoto's theory of stability:

Theorem 2.96 (Donaldson–Uhlenbeck–Yau). *Let E be an $\mathrm{SU}(r)-$ or a $\mathrm{PU}(r)-$ bundle over a* compact *Kähler manifold (Z, ω). There exists a one-to-one correspondence between gauge equivalence classes of HYM connections on E and isomorphism classes of polystable holomorphic bundles $\mathcal{E}$ whose underlying bundle is E.*

The following straightforward result realises HYM connections as a particular instance of $\Xi-$ASD instantons, for suitable choice of Ξ.

---

[7] $\mathcal{F}^* := \operatorname{Hom}(\mathcal{F}, \mathcal{O}_Z)$, where $\mathcal{O}_Z$ is the structure sheaf of Z.

Lemma 2.97. *Let E be an* $\mathrm{SU}(r)-$ *or a* $\mathrm{PU}(r)-$*bundle over a Kähler manifold* (Z^{2m}, ω) *of complex dimension* $m \geqslant 2$. *Consider the following closed* $(2m - 4)-$*form* Ξ *on* Z:

$$\Xi := \frac{\omega^{m-2}}{(m-2)!}.$$

Then, a connection $\nabla \in \mathcal{A}(E)$ *is a* $\Xi-$*ASD instanton if, and only if,* ∇ *is HYM.*

Finally, let $H = \mathrm{SU}(m) \subseteq \mathrm{SO}(2m)$, so that $N(H) = \mathrm{U}(m)$. If ω_0 is the standard Kähler form on $\mathbb{R}^{2m}$, then S. Salamon (1989, Chapter 3):

$$\Lambda^2(\mathbb{R}^{2m})^* = [[\Lambda^{2,0}]] \oplus [\Lambda^{1,1}_0] \oplus \langle \omega_0 \rangle,$$

where $[[\Lambda^{2,0}]] \otimes \mathbb{C} = \Lambda^{2,0} \oplus \Lambda^{0,2}$ and $[\Lambda^{1,1}_0] \otimes \mathbb{C} = \Lambda^{1,1}_0 := \ker(\Lambda_{\omega_0}|_{\Lambda^{1,1}})$. Furthermore, via the isomorphism $\mathfrak{so}(2m) \simeq \Lambda^2(\mathbb{R}^{2m})^*$, one has

$$\mathfrak{h} = \mathfrak{su}(m) \simeq [\Lambda^{1,1}_0].$$

It follows that, for $\mathrm{SU}(r)-$ or $\mathrm{PU}(r)-$bundles, the notions of $\mathrm{SU}(m)-$instantons (Definition 2.92) and HYM connections coincide.

2.3.2 G_2-instantons

This section is based on Walpuski (2013b, Chapter 1) and D. A. Salamon and Walpuski (2017).

Proposition 2.98 (D. A. Salamon and Walpuski (ibid., Theorem 8.4)). $\Lambda^2(\mathbb{R}^7)^*$ *decomposes orthogonally into*

$$\Lambda^2(\mathbb{R}^7)^* = \Lambda^2_7 \oplus \Lambda^2_{14},$$

where Λ^2_7 *and* Λ^2_{14} *are irreducible representations of* G_2, *with* $\dim \Lambda^2_d = d$, *given by*

$$\Lambda^2_7 := \{\alpha : *_{\phi_0}\alpha = 2\alpha\} = \{v \lrcorner \phi_0 : v \in \mathbb{R}^7\} \simeq \Lambda^1_7, \quad and$$

$$\Lambda^2_{14} := \{\alpha : *_{\phi_0}\alpha = -\alpha\} = \{\alpha : \alpha \wedge \psi_0 = 0\} \simeq \mathfrak{g}_2 \equiv \mathrm{Lie}(\mathrm{G}_2),$$

where $\psi_0 := *\phi_0$, *and the last isomorphism comes from the metric identification* $\Lambda^2(\mathbb{R}^7) \simeq \mathfrak{so}(7) \supseteq \mathfrak{g}_2$.

It follows that we have an analogous splitting of $\Lambda^2 T^* Y$ for every almost G_2-manifold (Y^7, ϕ). By slight abuse of notation, we will also denote the corresponding summands by Λ^2_d. Moreover, we can now extend for general compact Lie groups G the notion of G_2-instanton on $G-$bundles given in Definition 2.92 as follows.

Definition 2.99. Let (Y^7, g_ϕ) be a G_2-manifold and let E be a $G-$bundle over a Y, where G is a compact Lie group. A connection $\nabla \in \mathcal{A}(E)$ is called a G_2-**instanton** if ∇ is a $\phi-$ASD instanton (Definition 2.80).

Remark 2.100. In the above situation, suppose further that G is semi-simple. Then, by Proposition 2.98, a G_2-instanton $\nabla \in \mathcal{A}(E)$ is characterised by the following equivalent conditions:

(i) $*(\phi \wedge F_\nabla) = -F_\nabla$;

(ii) $F_\nabla \wedge \psi = 0$;

(iii) $\pi_7(F_\nabla) = 0$, for the orthogonal projection $\pi_7 \colon \Lambda^2 T^* Y \to \Lambda_7^2$;

(iv) F_∇ lies in $\mathfrak{g}_2 \otimes (\mathfrak{g}_E)_y \subseteq \Lambda^2 T_y^* Y \otimes (\mathfrak{g}_E)_y$, at each $y \in Y$. $\qquad \diamond$

Example 2.101. By Proposition 2.6 and Remark 2.100 (iv), the Levi-Civita connection D^{g_ϕ} of a connected G_2-manifold (Y^7, g_ϕ) is a G_2-instanton on the tangent bundle TY.

In the next example, we give an extension to the case $n = 7$ and $\Xi = \phi_0$ of Proposition 2.88.

Example 2.102 (G_2-instantons from ASD instantons). Let (Z^4, ω, Υ) be a Calabi–Yau $2-$fold and consider the G_2-manifold (Y^7, ϕ) of Example 2.30, where $Y^7 := \mathbb{R}^3 \times Z^4$ (resp. $Y^7 := T^3 \times Z^4$); write $\pi_Z \colon Y \to Z$ for the natural projection map. The following result relates $\mathbb{R}^3-$invariant (resp. T^3-invariant) G_2-instantons over Y with ASD connections over Z:

Proposition 2.103. *Let $E \to Z$ be a $G-$bundle with compact semi-simple structure group. A connection $I \in \mathcal{A}(E)$ is an ASD instanton if, and only if, $\nabla := \pi_Z^* I$ is a G_2-instanton.*

Proof. Note that

$$*(\phi \wedge F_\nabla) = *(\mathrm{d}x^{123} \wedge F_\nabla) = \pi_Z^*(*_Z F_I),$$

using $\omega \wedge F_\nabla = \mathrm{Re}(\Upsilon) \wedge F_\nabla = \mathrm{Im}(\Upsilon) \wedge F_\nabla = 0$, and the compatibility of ϕ with the product structures on Y. Since π_Z is a submersion and $F_\nabla = \pi_Z^* F_I$, the result follows. $\qquad \square$

Remark 2.104 (G_2-instanton equations in $\mathbb{R}^7$). Consider the model G_2-manifold $(\mathbb{R}^7, \phi_0)$ of Example 2.25. Let $\widetilde{E} \to \mathbb{R}^7$ be a (necessarily) trivial

G−bundle with compact semi-simple structure group, and let $\nabla \in \mathcal{A}(\widetilde{E})$. In Euclidean coordinates $x^1, \ldots, x^7$,

$$F_\nabla = \frac{1}{2} \sum F_{ij} \otimes dx^i \wedge dx^j, \quad F_{ij} : \mathbb{R}^7 \to \mathfrak{g}.$$

By Remark 2.100, ∇ is a G_2−instanton if, and only if, $F_\nabla \wedge \psi_0 = 0$, with $\psi_0 := *\phi_0$ given by

$$\psi_0 = dx^{4567} - dx^{1247} - dx^{1256} - dx^{2345} - dx^{2367} - dx^{3146} - dx^{3175}.$$

By a straightforward computation:

$$\begin{aligned}
F_\nabla \wedge \psi_0 = {} & (-F_{16} + F_{25} - F_{34}) \otimes dx^{123456} + (-F_{17} + F_{24} + F_{35}) \otimes dx^{123457} \\
& + (-F_{14} - F_{27} + F_{36}) \otimes dx^{123467} + (-F_{15} - F_{26} - F_{37}) \otimes dx^{123567} \\
& + (F_{12} - F_{47} - F_{56}) \otimes dx^{124567} + (F_{13} + F_{46} - F_{57}) \otimes dx^{134567} \\
& + (F_{23} - F_{45} - F_{67}) \otimes dx^{234567}.
\end{aligned}$$

Hence, ∇ is a G_2−instanton if, and only if,

$$\begin{cases}
F_{25} = F_{16} + F_{34}; & F_{12} = F_{47} + F_{56}; \\
F_{17} = F_{24} + F_{35}; & F_{57} = F_{13} + F_{46}; \\
F_{27}; & F_{23} = F_{45} + F_{67}; \\
F_{62} = F_{15} + F_{37}.
\end{cases} \tag{2.105}$$

Now write $\mathbb{R}^7 = \mathbb{R}^3 \oplus \mathbb{R}^4$ as in Remark 2.16. Denoting by $\pi : \mathbb{R}^7 \to \mathbb{R}^4$ the natural projection, suppose that $\widetilde{E} = \pi^* E$ is the pull-back of a G−bundle $E \to \mathbb{R}^4$ and that $\nabla = \pi^* I$, for some $I \in \mathcal{A}(E)$. Then, it follows explicitly from (2.105) and (1.63) that ∇ is a G_2−instanton if, and only if, I is an ASD instanton. Recalling from Example 2.70 that $\mathbb{R}^3 \times \{0\} \subseteq \mathbb{R}^7$ is ϕ_0−calibrated, this gives an explicit instance of Proposition 2.88 (for $n = 7$ and $\Xi = \phi_0$).

Example 2.106 (G_2−instantons from HYM-connections)**.** Let (Z^6, ω, Υ) be a Calabi–Yau 3−fold and consider the G_2−manifold (Y^7, ϕ) of Example 2.31, where $Y^7 := \mathbb{R} \times Z^6$ (resp. $Y^7 := S^1 \times Z^6$); denote by $\pi_Z : Y \to Z$ the natural projection map. The following result relates $\mathbb{R}$−invariant (resp. S^1−invariant) G_2−instantons over Y with HYM connections over Z (cf. Sá Earp (2015, Proposition 8) or Sá Earp and Walpuski (2015, Proposition 3.10)):

Proposition 2.107. *Let E be an* $\mathrm{SU}(r)-$ *or a* $\mathrm{PU}(r)-$*bundle over Z. A connection $\nabla \in \mathcal{A}(E)$ is HYM if, and only if, $\pi_Z^* \nabla$ is a G_2−instanton.*

Sketch of proof. The main point is to note that, in this context of a Calabi–Yau 3–fold, the HYM condition (2.94) is equivalent to

$$F_\nabla \wedge \mathrm{Im}(\Upsilon) = 0 \quad \text{and} \quad F_\nabla \wedge \omega \wedge \omega = 0.$$

The claim follows from

$$\psi = *(\mathrm{d}t \wedge \omega + \mathrm{Re}(\Upsilon)) = \frac{1}{2}\omega \wedge \omega - \mathrm{d}t \wedge \mathrm{Im}(\Upsilon),$$

since $\pi_Z^* \nabla$ is a G_2–instanton precisely when $F_{\pi_Z^* \nabla} \wedge \psi = 0$. $\qquad\square$

Remark 2.108. This basic result gives a way to obtain G_2–instantons on the ACyl building blocks $Y_\pm^7 := S^1 \times Z_\pm^6$ of the twisted connected sum construction, by solving the HYM problem on $Z_\pm^6$. This is indeed the stated motivation for the construction of HYM connections on such spaces (see below). $\qquad\diamond$

Over the past decade, non-trivial examples of G_2–instantons have gradually been constructed, in several contexts. Let us review some significant examples.

The first non-trivial construction appeared in Sá Earp (2009), initiating a project to construct G_2–instantons over twisted connected sums. A method to produce large amounts of examples, by means of the Hartshorne–Serre correspondence on 3-folds, was described in Jardim et al. (2017). A gluing theorem for such solutions, under suitable compatibility and transversality assumptions, was formulated in Sá Earp and Walpuski (2015), and explicit examples satisfying those conditions were further found in Menet, Nordström, and Sá Earp (2015) and Walpuski (2016). A thorough survey of this track can be found in Sá Earp (2018).

Along a different track, Walpuski (2013a,b) presented a method for constructing G_2–instantons over G_2–manifolds arising from Joyce's generalised Kummer construction Joyce (1996, 2000), providing some concrete examples with structure group SO(3).

More recently, Jacob and Walpuski (2018) proved an analogue of the Donaldson–Uhlenbeck–Yau theorem (cf. Theorem 2.96) for asymptotically cylindrical Kähler manifolds, handling reflexive sheaves. This provides examples of (singular) HYM connections over a certain class of complete noncompact Kähler manifolds, generalizing the result of Sá Earp (2015).

Another interesting construction of G_2–instantons, due to Clarke (2014), provides non-trivial examples on the trivial SU(2)–bundle over the total space of the spinor bundle $\mathbb{S}(S^3)$ of the round 3–sphere S^3, as in Bryant and S. Salamon (1989).

Finally, a recent exciting trend consists of constructions of G_2-instantons on spaces with symmetries, in particular cohomogeneity-one actions. Instances can be found in Ball and Oliveira (2019), Clarke and Oliveira (2019), and Lotay and Oliveira (2018).

Topological energy bounds from Chern–Weil theory.   Suppose (Y^7, g_ϕ) is a compact G_2−manifold and let E be an $SU(r)$−bundle over Y. For $\nabla \in \mathcal{A}(E)$, write $F_\nabla = F_\nabla^7 \oplus F_\nabla^{14}$ according to the decomposition of Λ^2 induced by $*_\phi$.
  Define the topological number

$$\kappa(E, [\phi]) := \langle c_2(E) \cup [\phi], [Y] \rangle .$$

Then we have:

$$
\begin{aligned}
8\pi^2 \kappa(E, [\phi]) &= \int_Y \mathrm{tr}(F_\nabla \wedge F_\nabla) \wedge \phi \\
&= -\langle F_\nabla, *(F_\nabla \wedge \phi)\rangle \\
&= -\langle F_\nabla^7 + F_\nabla^{14}, 2F_\nabla^7 - F_\nabla^{14}\rangle \\
&= -2\|F_\nabla^7\|_{L^2}^2 + \|F_\nabla^{14}\|_{L^2}^2
\end{aligned}
$$

On the other hand, $\mathcal{YM}(\nabla) = \|F_\nabla^7\|_{L^2}^2 + \|F_\nabla^{14}\|_{L^2}^2$. Therefore

$$\mathcal{YM}(\nabla) = 3\|F_\nabla^7\|_{L^2}^2 + 8\pi^2\kappa(E, [\phi]) = \frac{1}{2}\left(3\|F_\nabla^{14}\|_{L^2}^2 - 8\pi^2\kappa(E, [\phi])\right).$$

Hence, if $\kappa(E, [\phi]) > 0$ then G_2−instantons are absolute minima of $\mathcal{YM}$ attaining the topological energy bound $8\pi^2\kappa(E, [\phi])$; if $\kappa(E, [\phi]) < 0$ then E does not admit G_2−instantons at all.

2.3.3   Spin(7)−instantons

This section is based on Walpuski (2013b, Chapter 1) and D. A. Salamon and Walpuski (2017).

  The following result can be found in D. A. Salamon and Walpuski (ibid., p. 52, Theorem 9.5).

Proposition 2.109. $\Lambda^2(\mathbb{R}^8)^*$ *decomposes orthogonally into*

$$\Lambda^2(\mathbb{R}^8)^* = \Lambda_7^2 \oplus \Lambda_{21}^2,$$

where Λ_7^2 and Λ_{21}^2 are irreducible representations of Spin(7)*, with dim $\Lambda_d^2 = d$, given by*

$$\Lambda_7^2 := \{\alpha : *_{\Phi_0}\alpha = 3\alpha\}$$

$$\Lambda_{21}^2 := \{\alpha : *_{\Phi_0}\alpha = -\alpha\} \simeq \mathfrak{spin}(7),$$

where the last isomorphism comes from the metric identification $\Lambda^2(\mathbb{R}^8) \simeq \mathfrak{so}(8) \supseteq \mathfrak{spin}(7)$.

It follows that we have an analogous eigenspace decomposition of $\Lambda^2 T^* X$ with respect to $*_\Phi$ for every almost Spin(7)$-$manifold (X^8, Φ). By slight abuse of notation, we will also denote the corresponding summands by Λ_d^2.

In the light of the above result, we now extend for general compact Lie groups G the notion of Spin(7)$-$instanton on $G-$bundles given in Definition 2.92.

Definition 2.110. Let (X^8, Φ) be a Spin(7)$-$manifold and let E be a $G-$bundle over X where G is a compact Lie group. A connection $\nabla \in \mathcal{A}(E)$ is called a Spin(7)$-$**instanton** if ∇ is a $\Phi-$ASD instanton (cf. Definition 2.80).

Remark 2.111. In the above situation, suppose further that G is semi-simple. Thus, by Proposition 2.109, a connection $\nabla \in \mathcal{A}(E)$ is a Spin(7)$-$instanton precisely when one of the following equivalent conditions holds:

(i) $*(\Phi \wedge F_\nabla) = -F_\nabla$;

(ii) $\pi_7(F_\nabla) = 0$, where π_7 denotes the orthogonal projection from $\Lambda^2 T^* X$ to Λ_7^2;

(iii) The curvature tensor F_∇ lies in the subspace $\mathfrak{spin}(7) \otimes (\mathfrak{g}_E)_x \subseteq \Lambda^2 T_x^* X \otimes (\mathfrak{g}_E)_x$ at each $x \in X$. $\qquad\qquad\diamond$

Example 2.112. It follows from Proposition 2.6 and Remark 2.111 (iii) that if (X^8, g_Φ) is a connected Spin(7)$-$manifold, then D^{g_Φ} is a Spin(7)$-$instanton on TX.

Example 2.113 (Spin(7)$-$instantons from G_2-instantons)**.** Let (Y^7, ϕ) be a G_2-manifold and consider the associated Spin(7)$-$manifold (X^8, Φ) of Example 2.41, where $X^8 := \mathbb{R} \times Y^7$ (resp. $X^8 := S^1 \times Y^7$). The following result relates $\mathbb{R}-$invariant (resp. S^1-invariant) Spin(7)$-$instantons over X with G_2-instantons connections over Y:

Proposition 2.114. *Let E be a $G-$bundle over Y, where G is a compact semi-simple Lie group. A connection $\nabla \in \mathcal{A}(E)$ is a G_2-instanton if, and only if, $\pi_Y^* \nabla$ is a Spin(7)$-$instanton.*

Proof. By Remark 2.111, we know that ∇ is a G_2-instanton $\iff \phi \wedge F_\nabla = -*_Y F_\nabla \iff \psi \wedge F_\nabla = 0$. Thus, noting that

$$\begin{aligned}
*(\Phi \wedge F_{\pi_Y^* \nabla}) &= *(dt \wedge \pi_Y^*(\phi \wedge F_\nabla) + \pi_Y^*(\psi \wedge F_\nabla)) \\
&= \pi_Y^*(*_Y(\phi \wedge F_\nabla)) + *\pi_Y^*(\psi \wedge F_\nabla),
\end{aligned}$$

we are done. $\qquad\qquad\square$

Example 2.115. Spin(7)$-$instantons were the subject of Lewis' Ph.D. thesis, Lewis (1999). He constructs a non-trivial example on a SU(2)$-$bundle over a particular compact Riemannian 8$-$manifold, obtained by Joyce (1996), with holonomy exactly Spin(7). More recently, a construction for Spin(7)$-$instantons on 8$-$manifolds arising from Joyce (1999) was given by Tanaka (2012), and Walpuski (2017b) proved an existence theorem that applies to the construction of Spin(7)$-$instantons on Spin(7)$-$manifolds with suitable local K3 Cayley fibrations and in particular recovers the example constructed by Lewis. Moreover, Clarke (2014) constructed a (symmetric) Spin(7)$-$instanton with structure group SU(2) on the Bryant–Salamon Bryant and S. Salamon (1989) negative spinor bundle $\mathbb{S}^-(S^4)$, which is smooth away from the Cayley base (zero section) S^4, and blows up along the later.

Complex ASD instantons.   In this brief paragraph, we introduce the notion of complex ASD instanton over Calabi–Yau 4$-$folds and realise it as a particular instance of the notion of Spin(7)$-$instanton. Complex ASD instantons and its underlying 'complex gauge theory' was notably studied by R. Thomas in his Ph.D. thesis, Thomas (1997); see also Donaldson and Thomas (1998).

Let (Z^8, ω, Υ) be a Calabi–Yau 4$-$fold and consider the following operator:

$$*_\Upsilon : \Omega^{0,p}(Z) \to \Omega^{0,4-p}(Z)$$
$$\omega \mapsto \overline{*}(\omega \wedge \Upsilon),$$

where $\overline{*} : \Lambda^{p,q} T^* Z \to \Lambda^{n-p,n-q} T^* Z$ is the usual anti-linear Hodge star operator on Kähler manifolds. It follows that $*_\Upsilon$ gives an endomorphism

$$*_\Upsilon : \Omega^{0,2} \to \Omega^{0,2}$$

which is self-adjoint and squares to the identity, splitting $\Omega^{0,2}$ orthogonally into *real* subspaces $\Omega^{0,2}_\pm$ corresponding to the eigenvalues ± 1, in complete analogy with the familiar real 4$-$dimensional case.

Definition 2.116. A connection $\nabla \in \mathcal{A}(E)$ on an SU(r)$-$ or a PU(r)$-$bundle E over Z^8 is called a **complex ASD instanton** if

$$*_\Upsilon \, F_\nabla^{0,2} = -F_\nabla^{0,2}. \tag{2.117}$$

We can fit this notion into the context of $\Xi-$ASD instantons as follows. Consider on (Z^8, ω, Υ) the natural Spin(7)$-$structure Φ of Example 2.42. Then we have:

Lemma 2.118. *Let E be an* $\mathrm{SU}(r)-$ *or a* $\mathrm{PU}(r)-$*bundle over* Z^8 *and let* $\nabla \in \mathcal{A}(E)$. *Then* ∇ *is a complex ASD instanton if, and only if, ∇ is a* $\mathrm{Spin}(7)-$*instanton with respect to Φ.*

The proof of this lemma is just a matter of unraveling the definitions and taking account of bi-degrees.

Topological energy bounds from Chern–Weil theory. Suppose (X^8, Φ) is a compact $\mathrm{Spin}(7)-$manifold and let E be a $\mathrm{SU}(r)-$bundle over X. Let $\nabla \in \mathcal{A}(E)$ and write $F_\nabla = F_\nabla^7 \oplus F_\nabla^{21}$ according to the decomposition of Λ^2 induced by $*_\Phi$. Define the topological number

$$\kappa(E, [\Phi]) := \langle c_2(E) \cup [\Phi], [X] \rangle .$$

Then

$$
\begin{aligned}
8\pi^2 \kappa(E, [\Phi]) &= \int_X \mathrm{tr}(F_\nabla \wedge F_\nabla) \wedge \Phi \\
&= -\langle F_\nabla, *(F_\nabla \wedge \Phi) \rangle \\
&= -\langle F_\nabla^7 + F_\nabla^{21}, 3F_\nabla^7 - F_\nabla^{21} \rangle \\
&= -3\|F_\nabla^7\|_{L^2}^2 + \|F_\nabla^{21}\|_{L^2}^2
\end{aligned}
$$

On the other hand, $\mathcal{YM}(\nabla) = \|F_\nabla^7\|_{L^2}^2 + \|F_\nabla^{21}\|_{L^2}^2$. Therefore

$$\mathcal{YM}(\nabla) = 4\|F_\nabla^7\|_{L^2}^2 + 8\pi^2 \kappa(E, [\Phi]) = \frac{1}{3}\left(4\|F_\nabla^{21}\|_{L^2}^2 - 8\pi^2 \kappa(E, [\Phi])\right).$$

Hence, if $\kappa(E, [\Phi]) > 0$ then $\mathrm{Spin}(7)-$instantons are the absolute minima of $\mathcal{YM}$, which attains the topological energy bound $8\pi^2 \kappa(E, [\Phi])$; if $\kappa(E, [\Phi]) < 0$ then E does not admit $\mathrm{Spin}(7)-$instantons at all.

3
Analytical aspects of Yang–Mills connections

In this chapter we shall be interested in the analytical study of the weak-convergence and regularity theory of Yang–Mills connections in dimensions higher than (or equal to) four, following the seminal works of Uhlenbeck (1982a,b), Price (1983), Nakajima (1988) and Tian (2000).

Section 3.1 briiefly surveys weak and strong Uhlenbeck compactness results for connections with uniform L^p-bounds on curvature, where $1 < p < \infty$, $2p > n$ (Theorems 3.2 and 3.7). The section ends with a consequent compactness result for Yang–Mills connections with locally uniformly bounded curvatures (modulo passing to a subsequence) allowing for noncompact base manifolds. Thenceforth we leave the general setting of arbitrary dimensions and consider only higher dimensional base manifolds.

In Section 3.2 we deduce Price's monotonicity formula for Yang–Mills fields (Theorem 3.24), as well as its interesting corollary implying that there is no non-flat Yang–Mills connection with finite L^2-energy over the standard (flat) Euclidean space $\mathbb{R}^n$ for $n \geq 5$. Next, in Section 3.3, we derive a local estimate, due to Uhlenbeck and Nakajima, for the $L^\infty-$norm of Yang–Mills fields with sufficiently small normalized L^2-norm on small geodesic balls (Theorem 3.33). Then in Section 3.4 we derive a noncompactness phenomenon along sets of Hausdorff codimension at least four, for general sequences of Yang–Mills connections with uniformly L^2−bounded curvatures (Theorem 3.48).

Convention 3.1. Throughout this chapter, unless otherwise stated, (M, g) denotes a connected, oriented, Riemannian n−manifold, and E denotes a G−bundle over M, where G is a compact Lie group.

3.1   Uhlenbeck's compactness theorems

In her seminal paper Uhlenbeck (1982a), Uhlenbeck proved the local existence of the so-called Coulomb gauges for connections with local $L^{n/2}$−norm of the curvature sufficiently small. In particular, this enabled her to prove a global weak compactness theorem for arbitrary fields with bounded L^p−norm, for some $p > n/2$.

Let us now review the so-called compactness results of Uhlenbeck, following closely the excellent book by Wehrheim (2004). Our exposition will in fact be rather sketchy, because the results we recall here require a fair amount of background outside of our intended scope. This section is intended only to organise, for future reference, the main ideas of some important compactness results.

Weak Uhlenbeck compactness.   *Unless otherwise stated, we suppose our base manifold M to be* compact, *with (possibly empty) boundary.*
Recall from Proposition 1.26 that for $1 < p < \infty$ such that $p > \frac{n}{2}$, the space of $W^{2,p}$ gauge transformations $\mathcal{G}^{2,p}(E)$ forms a topological group (with respect to composition) which acts continuously on the space of $W^{1,p}$ connections $\mathcal{A}^{1,p}(E)$. In particular, we may consider the topological quotient $\mathcal{M}^p := \mathcal{A}^{1,p}(E)/\mathcal{G}^{2,p}(E)$. In this context, Uhlenbeck's weak compactness theorem asserts the weak compactness of subsets of the form $\{[\nabla] \in \mathcal{M}^p : \|F_\nabla\|_p \leqslant \Lambda\}$, for any constant $\Lambda > 0$. Indeed, we can state it as follows (cf. Uhlenbeck (1982a, Theorem 1.5) and Wehrheim (2004, Theorem 7.1, p. 108)):

Theorem 3.2 (Uhlenbeck). *Suppose $1 < p < \infty$ is such that $2p > n$. Let $\{\nabla_i\} \subseteq \mathcal{A}^{1,p}(E)$ be a sequence of connections such that $\|F_{\nabla_i}\|_p$ is uniformly bounded. Then, after passing to a subsequence, there exist gauge transformations $g_i \in \mathcal{G}^{2,p}(E)$ such that $g_i^* \nabla_i$ converges weakly in $\mathcal{A}^{1,p}(E)$.*

The main step in the proof of this weak compactness theorem is to show that 'Coulomb gauges' exist over small trivializing neighborhoods $U \subseteq M$ of E. In a fixed local trivialization $E|_U \simeq U \times \mathbb{K}^r$, note that the spaces $\mathcal{A}^{1,p}(E|_U)$ and $\mathcal{G}^{2,p}(E|_U)$ are represented, respectively, by $W^{1,p}(U, T^*U \otimes \mathfrak{g})$ and

$W^{2,p}(U,G)$. In the following theorem Wehrheim (ibid., p. 91, Theorem 6.1), for $A \in W^{1,p}(U, T^*U \otimes \mathfrak{g})$ and $g \in W^{2,p}(U,G)$, we use the notations:

$$F_A := \mathrm{d}A + A \wedge A,$$

$$\mathcal{E}^q(A) := \|F_A\|^q_{L^q(U)}, \text{ and}$$

$$g^*A := g^{-1}Ag + g^{-1}\mathrm{d}g.$$

Theorem 3.3 (Local Coulomb gauge). *Let* $1 < q \leqslant p < \infty$ *be such that* $q \geqslant \frac{n}{2}$, $p > \frac{n}{2}$ *and, in case* $q < n$, *assume in addition* $p \leqslant \frac{nq}{n-q}$. *Then there exist constants* $\kappa_0 \geqslant 0$ *and* $\gamma_0 > 0$ *such that the following holds: for each* $x \in M$, *given any neighborhood* $\tilde{U}$ *of* x *in* M *there exists a geodesic ball* $U \subseteq \tilde{U}$ *around* x *such that for every* $A \in W^{1,p}(U, T^*U \otimes \mathfrak{g})$ *with* $\mathcal{E}^q(A) \leqslant \gamma_0$ *there exists a gauge transformation* $g \in W^{2,p}(U,G)$ *such that* g^*A *is in* Coulomb *gauge, i.e. the following holds:*

(i) $\mathrm{d}^*(g^*A) = 0.$

(ii) $*(g^*A)|_{\partial U} = 0.$

(iii) $\|g^*A\|_{W^{1,q}(U)} \leqslant \kappa_0\|F_A\|_{L^q(U)}.$

(iv) $\|g^*A\|_{W^{1,p}(U)} \leqslant \kappa_0\|F_A\|_{L^p(U)}.$

Remark 3.4. In Wehrheim (ibid., Remark 6.2 (a), p. 91), it is shown that the theorem also holds for the case $q = p = \frac{n}{2}$ provided $n \geqslant 3$. In this way, the above theorem is a generalization of Uhlenbeck's original version Uhlenbeck (1982a, Theorem 1.3), which corresponds to the case $q = \frac{n}{2}$ and $n \geqslant p \geqslant \frac{n}{2}$ of the above result[1]. $\Diamond$

The above theorem is proved by first solving the boundary value problem given by (i)-(ii) and then deducing (iii)-(iv) from *a priori* bounds. Such *a priori* bounds are given by the following regularity result.

Theorem 3.5 (Regularity for $\mathrm{d} \oplus \mathrm{d}^*$). *Let* (M, g) *be a compact* n−*manifold with (possibly empty) boundary and let* $1 < p < \infty$. *Then, there exists a constant* $C > 0$ *such that for every* $A \in W^{1,p}(M, T^*M)$ *satisfying* $*A|_{\partial M} = 0$ *we have*

$$\|A\|_{1,p} \leqslant C \left(\|\mathrm{d}A\|_p + \|\mathrm{d}^*A\|_p + \|A\|_p\right).$$

Moreover, if in addition $H^1(M, \mathbb{R}) = 0$, *then we can drop the* $\|A\|_p$ *term on the RHS of the above estimate.*

---

[1] *"(...) it seems that in order to obtain a* $W^{1,p}$−*control in (iv) for* $p > n$, *one needs small energy for* $q > \frac{n}{2}$." Wehrheim (2004, Remark 6.2 (c), p. 92).

In fact, given the local aspect of Theorem 3.3, one actually first reduces the general setting to model cases. Given $x \in M$ and $\delta > 0$, according to weather $x \in \mathrm{int}(M)$ or $x \in \partial M$, we may find an appropriate domain $B \subseteq \mathbb{R}^n$, a constant $\sigma \in]0, 1]$ and a chart $\psi_\sigma : B \to M$ centered at x such that[2]

$$\|\sigma^{-2}\psi_\sigma^* g - \mathbb{1}\|_{2,\infty} \leqslant \delta.$$

Thus, by working in such special local coordinates, it suffices to prove the local Coulomb gauge theorem when $M = B$ is equipped with a smooth metric g satisfying $\|g-\mathbb{1}\|_{2,\infty} \leqslant \delta$, for some sufficiently small $\delta > 0$, and then examine the effect of rescaling the metric.

A key property of a local Coulomb gauge, explored in the proof of Theorem 3.2, is that in such a gauge we can pass the uniform L^p–control on the curvatures F_{∇_i} to a uniform $W^{1,p}$–control on the connections matrices (cf. Theorem 3.3 (iv)). Thus, by the reflexiveness of the Sobolev spaces $W^{1,p}$, in each local Coulomb gauge we can extract a weakly $W^{1,p}$–convergent subsequence of the connections ∇_i (as a consequence of the Banach–Alaoglu theorem - see Appendix B). Ultimately, one has to patch together these local gauges in a suitable way to complete the proof of the weak compactness theorem.

For the sake of completeness, we state below a general patching result. First, fix in G the natural bi-invariant Riemannian metric induced by $\langle \cdot, \cdot \rangle_{\mathfrak{g}}$ and let d_G denote the Riemannian distance function in G with respect to such metric. Next, let $\Delta_{\exp} > 0$ be the radius of a convex geodesic ball $B_{\Delta_{\exp}}(1_G) \subseteq G$ centered at 1_G, such that the following holds:

1. The exponential map exp restricted to $B_{\Delta_{\exp}}(0) \subseteq \mathfrak{g}$ is a diffeomorphism onto $B_{\Delta_{\exp}}(1_G)$.

2. For all $g, h \in B_{\Delta_{\exp}}(1_G)$ there exists a unique minimal geodesic from g to h and this lies within $B_{\Delta_{\exp}}(1_G)$.

Lemma 3.6 (Wehrheim (2004, p. 111, Lemma 7.2)). *Let M be an n–manifold and let $p > \frac{n}{2}$. Suppose $\{U_\alpha\}$ is a locally finite open covering of M by precompact sets U_α, where α runs a countable index set I. Then there exist open subsets $V_\alpha \subseteq U_\alpha$ still covering M such that the following holds.*

(i) Let $k \in \mathbb{N}$ and let $g_{\alpha\beta}, h_{\alpha\beta} \in \mathcal{G}^{k+1,p}(U_\alpha \cap U_\beta)$ be two sets of transition functions satisfying the cocycle conditions and

$$d_G(g_{\alpha\beta}, h_{\alpha\beta}) \leqslant \Delta_{exp}, \quad \forall \alpha, \beta \in I.$$

---

[2]Here we are identifying $\psi_\sigma^* g$ with its matrix representation in canonical coordinates, and $\mathbb{1}$ denotes the identity matrix of order n.

Then there exist local gauge transformations $h_\alpha \in \mathcal{G}^{k+1,p}(V_\alpha)$ for all $\alpha \in I$ such that on all intersections $V_\alpha \cap V_\beta$ we have

$$h_\alpha^{-1} h_{\alpha\beta} h_\beta = g_{\alpha\beta}.$$

(ii) Let the $h_{\alpha\beta}$ in (i) run through a sequence $h_{\alpha\beta}^i$ of sets of transition functions such that $g_{\alpha\beta}, h_{\alpha\beta}^i \in \mathcal{G}^{k+1,p}(U_\alpha \cap U_\beta)$ for all $k < K$, where $K \geqslant 2$ is an integer or $K = \infty$. Assume that for every $\alpha, \beta \in I$ and $k < K$ there is a uniform bound on $\|(h_{\alpha\beta}^i)^{-1} \mathrm{d}h_{\alpha\beta}^i\|_{W^{k,p}(U_\alpha \cap U_\beta)}$.

Then the gauge transformations h_α^i in (i) are constructed in such a way that for every $\alpha \in I$ and $k < K$ they satisfy $h_\alpha^i \in \mathcal{G}^{k+1,p}(V_\alpha)$ and

$$\sup_{i \in \mathbb{N}} \|(h_\alpha^i)^{-1} \mathrm{d}h_\alpha^i\|_{W^{k,p}(V_\alpha)} < \infty.$$

Strong Uhlenbeck compactness. *In this paragraph, unless otherwise stated, we suppose our base manifold M is a* compact *manifold with (possibly empty) boundary*[3].

Besides the generality and power of the weak compactness theorem, it can be greatly improved when we restrict ourselves to sequences of (weak) *Yang–Mills* connections.

Theorem 3.7 (Strong Uhlenbeck compactness). *Let $1 < p < \infty$ be such that $p > \frac{n}{2}$ and, in case $n = 2$, assume in addition $p \geqslant \frac{4}{3}$. Let $\{\nabla_i\} \subseteq \mathcal{A}^{1,p}(E)$ be a sequence of weak Yang–Mills connections such that $\|F_{\nabla_i}\|_p$ is uniformly bounded. Then, after passing to a subsequence, there exist gauge transformations $g_i \in \mathcal{G}^{2,p}(E)$ such that $\{g_i^* \nabla_i\} \subseteq \mathcal{A}(E)$ is a sequence of smooth Yang–Mills connections that converges to a smooth Yang–Mills connection $\nabla \in \mathcal{A}(E)$ in C^∞–topology.*

The key result in the proof of the strong Uhlenbeck compactness is the existence of global relative Coulomb gauges.

Theorem 3.8 (Relative Coulomb gauge). *Let $1 < p \leqslant q < \infty$ be such that $p > \frac{n}{2}$ and $\frac{1}{n} > \frac{1}{q} > \frac{1}{p} - \frac{1}{n}$. Fix a reference connection $\nabla_0 \in \mathcal{A}^{1,p}$ and let a constant $c_0 > 0$ be given. Then there exist constants $\delta > 0$ and $C > 0$ such that the following holds. For every $\nabla \in \mathcal{A}^{1,p}$ with*

$$\|\nabla - \nabla_0\|_q \leqslant \delta \quad \text{and} \quad \|\nabla - \nabla_0\|_{1,p} \leqslant c_0,$$

there exists a gauge transformation $g \in \mathcal{G}^{2,p}(E)$ such that

---

[3] It is important to note that, in such context, we consider the Yang–Mills equation $\mathrm{d}_\nabla^* F_\nabla = 0$ with the boundary condition $*F_\nabla|_{\partial M} = 0$ (see the footnote of number 18 in Chapter 1).

(i) $d^*_{\nabla_0}(g^*\nabla - \nabla_0) = 0.$

(ii) $\|g^*\nabla - \nabla_0\|_q \leqslant C\|\nabla - \nabla_0\|_q.$

(iii) $\|g^*\nabla - \nabla_0\|_{1,p} \leqslant C\|\nabla - \nabla_0\|_{1,p}.$

By iteration of regularity results, one of the consequences of the relative Coulomb gauge theorem is the following:

Theorem 3.9 (Regularity of weak Yang–Mills connections). *Let* $1 < p < \infty$ *be such that* $p > \frac{n}{2}$ *and, in case* $n = 2$*, assume in addition* $p \geqslant \frac{4}{3}$*. Then, for every weak Yang–Mills connection* $\nabla \in \mathcal{A}^{1,p}(E)$ *there exists a gauge transformation* $g \in \mathcal{G}^{2,p}(E)$ *such that* $g^*\nabla$ *is a smooth connection.*

Once one proves such results, the strong Uhlenbeck compactness (Theorem 3.7) is basically reduced to the weak Uhlenbeck compactness (Theorem 3.2) without using a further patching argument[4]. The argument, due to Dietmar Salamon, can be outlined as follows (cf. Wehrheim (2004, p. 153)). First, by the weak compactness theorem, after passing to a subsequence, we may find gauge transformations $g_i \in \mathcal{G}^{2,p}(E)$ such that $g_i^*\nabla_i$ converges in the weak $W^{1,p}$–topology to some $\nabla \in \mathcal{A}^{1,p}(E)$. It can be shown that ∇ also is a weak Yang–Mills connection[5], so that after a gauge transformation we can suppose it is smooth (by Theorem 3.9). Moreover, after passing to a further subsequence, we can suppose that $\|\nabla_i - \nabla\|_{1,p}$ is bounded and that, for a suitable $1 < p \leqslant q < \infty$ such that the Sobolev embedding $W^{1,p} \hookrightarrow L^q$ is compact, the ∇_i converges to ∇ in the L^q–norm. Finally, one puts the connections ∇_i in relative Coulomb gauge with respect ∇ (Theorem 3.8). The C^∞–convergence follows from the fact that the Yang–Mills equation together with the relative Coulomb gauge condition form an elliptic system, thus provide uniform bounds on all $W^{k,p}$–norms of the connections, and the compactness then follows from the compact Sobolev embeddings (Theorem B.13 (ii)).

Compactness theorem for smooth Yang–Mills connections.   We finish this section with a compactness result for smooth Yang–Mills connections whose proof follows the same line of argument of the proof of the weak Uhlenbeck compactness.

---

[4]The 'standard' proof of the strong compactness theorem essentially follows the same line of argument of the proof of the weak compactness: one finds local Coulomb gauges in which one has convergent subsequences and then use a patching construction to obtain global gauges (see e.g. Donaldson and Kronheimer (1990, §4.4.2–4.4.3)).

[5]Here one needs $p > \frac{4}{3}$ in case $n = 2$.

Theorem 3.10. *Let $\{\nabla_i\} \subseteq \mathcal{A}(E)$ be a sequence of smooth Yang–Mills connections with the following property. For each $x \in M$, there exist a neighborhood U of x and a subsequence $\{i'\} \subseteq \{i\}$ such that $|F_{\nabla_{i'}}|$ is uniformly bounded on U. Then there exist a single subsequence $\{i''\} \subseteq \{i\}$, a sequence of smooth gauge transformations $\{g_{i''}\} \subseteq \mathcal{G}(E)$ and a smooth Yang–Mills connection $\nabla \in \mathcal{A}(E)$ such that the sequence $g_{i''}^{*}\nabla_{i''}$ converges to ∇ in C^{∞}–topology on compact subsets of M.*

The proof of Theorem 3.10 uses the local Coulomb gauge Theorem 3.3, elliptic regularity, Arzelà-Ascoli and the following standard patching argument from Donaldson and Kronheimer (1990, Corollary 4.4.8, p. 160):

Proposition 3.11. *Let $\{\nabla_i\} \subseteq \mathcal{A}(E)$ be a sequence of smooth connections with the following property. For each $x \in M$, there exist a neighborhood U of x, a subsequence $\{i'\} \subseteq \{i\}$, and gauge transformations $\{g_{i'}\} \subseteq \mathcal{G}(E|_U)$ such that $g_{i'}^{*}\nabla_{i'}$ is convergent in C^{∞}–topology on compact sets in U. Then there exist a single subsequence $\{i''\}$ and smooth gauge transformations $g_{i''} \in \mathcal{G}(E)$ such that $g_{i''}^{*}\nabla_{i''}$ converges in C^{∞}–topology on compact sets over all of M.*

One of the goals of the next two sections is to achieve a compactness theorem for Yang–Mills connections in dimension $n \geq 4$ assuming only that the L^2–norm of the curvatures are uniformly bounded. (Recall from Proposition 2.84 that we have *a priori* L^2–energy bound for Ξ–ASD instantons, provided G is semi-simple.) One then needs to use *a priori* estimates to bound the pointwise norm of curvature and the convergence will be possible only away from a blow-up set where the L^2–energy of the sequence concentrates (cf. Section 3.4, Theorem 3.48).

3.2 Price's monotonicity formula

For the rest of this chapter, we assume that $n := \dim M \geq 4$.

Price's monotonicity formula Price (1983) is a key result in the analysis of Yang–Mills fields in higher dimensions. In particular, it allows normalized L^2–energy estimates on balls to pass down to smaller balls. Following Tian (2000, §2.1), in this section we derive (a slightly modified version of) Price's result that will play a pivotal role in the rest of this work.

A variational formula. We start by deriving a variational formula for the Yang–Mills action along vector fields with compact support (cf. Tian (ibid., pp. 208-210) and Price (1983, pp. 141-146)).

Let $X \in \mathfrak{X}(M)$ be a vector field with compact support and denote by $\{\phi_t\}$ the *flow* of X, i.e. the induced 1–parameter family of diffeomorphisms $\phi_t : M \to M$. Note that each ϕ_t restricts to the identity map outside the support of X.

Given a smooth connection $\nabla \in \mathcal{A}(E)$ with $\mathcal{YM}(\nabla) < \infty$, the flow of X induces a compactly supported variation $\{\nabla_t\}$ of ∇ as follows. Denote by $P_t : E_x \to E_{\phi_t(x)}$ the parallel transport, with respect to ∇, along the path $\{\phi_s(x)\}_{0 \leqslant s \leqslant t}$ (cf. Section 1.2). Note that P_t acting on sections of E gives rise to sections of the induced bundle $\phi_t^* E$, in such a way that

$$P_t(fs) = (\phi_t^* f)P_t s, \tag{3.12}$$

for each $f \in C^\infty(M)$ and $s \in \Gamma(E)$.

For each t, consider the pull-back connection $\phi_t^* \nabla$ on $\phi_t^* E \to M$ (see 1.1) and define:

$$\nabla_t s := (P_t)^{-1}\left(\phi_t^* \nabla\right)(P_t s), \quad \text{for each } s \in \Gamma(E).$$

It is clear that $\nabla_0 = \nabla$. To verify that each ∇_t indeed defines a connection, note first that linearity follows from the fact that P_t (therefore $(P_t)^{-1}$) and $\phi_t^* \nabla$ are linear maps. Moreover, for every $f \in C^\infty(M)$ and $s \in \Gamma(E)$ we have:

$$\begin{aligned}
\nabla_t(fs) &= (P_t)^{-1}\left(\phi_t^* \nabla\right)(P_t(fs)) \\
&= (P_t)^{-1}\left(\phi_t^* \nabla\right)((\phi_t^* f)P_t s) \quad \text{(by (3.12))} \\
&= (P_t)^{-1}\left(\mathrm{d}(\phi_t^* f) \otimes (P_t s) + \phi_t^* f(\phi_t^* \nabla)(P_t s)\right) \quad (\phi_t^* \nabla \text{ is a connection}) \\
&= \phi_{-t}^*\left(\mathrm{d}(\phi_t^* f)\right) \otimes s + \phi_{-t}^*\left(\phi_t^* f\right)(P_t)^{-1}\left((\phi_t^* \nabla)(P_t s)\right) \quad \text{(by (3.12))} \\
&= \mathrm{d}f \otimes s + f\nabla_t s.
\end{aligned}$$

Now, for each t, we have $F_{\phi_t^* \nabla} = \phi_t^* F_\nabla$, so that the associated curvature F_{∇_t} is given by

$$F_{\nabla_t} = (P_t)^{-1} \circ (\phi_t^* F_\nabla) \circ P_t.$$

Therefore:

$$F_{\nabla_t}(Y, Z) = (P_t)^{-1} \cdot F_\nabla(\mathrm{d}\phi_t(Y), \mathrm{d}\phi_t(Z)) \cdot (P_t), \quad \forall Y, Z \in \mathfrak{X}(M). \tag{3.13}$$

We now wish to calculate $\dfrac{\mathrm{d}}{\mathrm{d}t}\mathcal{YM}(\nabla_t)\Big|_{t=0}$. Given a local orthonormal frame $\{e_i\}$ of TM, we have

$$|F_{\nabla_t}|^2(x) = \sum_{i,j}|F_{\nabla_t}(e_i(x), e_j(x))|_{\mathfrak{g}}^2$$

$$= \sum_{i,j}|F_\nabla(\mathrm{d}\phi_t(e_i(x)), \mathrm{d}\phi_t(e_j(x)))|_{\mathfrak{g}}^2 \quad \text{(by (3.13) and Adinvariance)}.$$

Thus, we can write[6]

$$\mathcal{YM}(\nabla_t) = \int_M \sum_{i,j} |F_\nabla(\mathrm{d}\phi_t(e_i(x)), \mathrm{d}\phi_t(e_j(x)))|_{\mathfrak{g}}^2 \mathrm{d}V_g(x).$$

By changing variables, we get

$$\mathcal{YM}(\nabla_t) = \int_M \sum_{i,j} |F_\nabla(\mathrm{d}\phi_t(e_i(\phi_t^{-1}(x))), \mathrm{d}\phi_t(e_j(\phi_t^{-1}(x))))|_{\mathfrak{g}}^2 \mathrm{Jac}(\phi_t^{-1})\mathrm{d}V_g(\phi_t^{-1}(x)).$$

Now note that

$$\frac{\mathrm{d}}{\mathrm{d}t}\left(\mathrm{d}\phi_t(e_j(\phi_t^{-1}(x)))\right)\Big|_{t=0} = -[X, e_i](x),$$

and

$$\begin{aligned}
\frac{\mathrm{d}}{\mathrm{d}t}\left(\mathrm{Jac}(\phi_t^{-1})(x)\mathrm{d}V_g(\phi_t^{-1}(x))\right)\Big|_{t=0} &= \frac{\mathrm{d}}{\mathrm{d}t}\left((\phi_t^{-1})^*\mathrm{d}V_g\right)(x)\Big|_{t=0} \\
&= \left(\mathcal{L}_{-X}\mathrm{d}V_g\right)(x) \\
&= -(\mathrm{div}\, X)(x)\mathrm{d}V_g(x).
\end{aligned}$$

Thus, the Leibniz rule and the chain rule give (cf.Tian (2000, pp. 209-210)):

$$\frac{\mathrm{d}}{\mathrm{d}t}\mathcal{YM}(\nabla_t)\Big|_{t=0} = -\int_M \left(|F_\nabla|^2\mathrm{div}\, X + 4\sum_{i,j=1}^n \langle F_\nabla([X, e_i], e_j), F_\nabla(e_i, e_j)\rangle_{\mathfrak{g}}\right)\mathrm{d}V_g.$$

If D denotes the Levi-Civita connection of (M, g), note that we can write

$$\begin{aligned}
&\sum_{i,j=1}^n \langle F_\nabla([X, e_i], e_j), F_\nabla(e_i, e_j)\rangle_{\mathfrak{g}} \\
&= -\sum_{i,j=1}^n \left(\langle F_\nabla(D_{e_i}X, e_j), F_\nabla(e_i, e_j)\rangle_{\mathfrak{g}} - \langle F_\nabla(D_X e_i, e_j), F_\nabla(e_i, e_j)\rangle_{\mathfrak{g}}\right) \\
&= -\sum_{i,j=1}^n \left(\langle F_\nabla(D_{e_i}X, e_j), F_\nabla(e_i, e_j)\rangle_{\mathfrak{g}} - \sum_{k=1}^n g(D_X e_i, e_k)\langle F_\nabla(e_k, e_j), F_\nabla(e_i, e_j)\rangle_{\mathfrak{g}}\right).
\end{aligned}$$

---

[6]Note that we can always cover M with open subsets over which TM trivializes by means of orthonormal frames - the tangent bundle of a Riemannian manifold is an $O(n)$−bundle; pick a partition of unity subordinate to such a cover to localize the integrand.

Further, since D is compatible with g,

$$\sum_{i,j,k=1}^{n} g(D_X e_i, e_k)\langle F_\nabla(e_k, e_j), F_\nabla(e_i, e_j)\rangle_{\mathfrak{g}}$$

$$= -\sum_{i,j,k=1}^{n} g(D_X e_k, e_i)\langle F_\nabla(e_k, e_j), F_\nabla(e_i, e_j)\rangle_{\mathfrak{g}}$$

$$= -\sum_{i,j,k=1}^{n} g(D_X e_k, e_i)\langle F_\nabla(e_i, e_j), F_\nabla(e_k, e_j)\rangle_{\mathfrak{g}} \quad (\text{symmetry of } \langle \cdot, \cdot \rangle_{\mathfrak{g}})$$

$$= -\sum_{i,j,k=1}^{n} g(D_X e_i, e_k)\langle F_\nabla(e_k, e_j), F_\nabla(e_i, e_j)\rangle_{\mathfrak{g}}. \quad (\text{interchanging } i \text{ and } k)$$

So we conclude that

$$\sum_{i,j=1}^{n} \langle F_\nabla([X, e_i], e_j), F_\nabla(e_i, e_j)\rangle_{\mathfrak{g}} = -\sum_{i,j=1}^{n} \langle F_\nabla(D_{e_i} X, e_j), F_\nabla(e_i, e_j)\rangle_{\mathfrak{g}}.$$

Summing up the above calculations, we have the following formula for the *first variation of $\mathcal{YM}$ along X*:

$$\frac{\mathrm{d}}{\mathrm{d}t}\mathcal{YM}(\nabla_t)\Big|_{t=0} =$$

$$-\int_M \left(|F_\nabla|^2 \mathrm{div}\, X - 4\sum_{i,j=1}^{n} \langle F_\nabla(D_{e_i} X, e_j), F_\nabla(e_i, e_j)\rangle_{\mathfrak{g}} \right) \mathrm{d}V_g. \quad (3.14)$$

If ∇ is a Yang–Mills connection, recalling Proposition 1.52, we deduce the variational formula:

$$\int_M \left(|F_\nabla|^2 \mathrm{div}\, X - 4\sum_{i,j=1}^{n} \langle F_\nabla(D_{e_i} X, e_j, F_\nabla(e_i, e_j))\rangle \right) \mathrm{d}V_g = 0. \quad (3.15)$$

We shall see that this stationary condition turns out to be the main ingredient in the proof of Price's monotonicity.

A word on notation. Henceforth, we will use the following notations concerning any (connected) Riemannian manifold (M, g):

- d_g: Riemannian distance function on (M, g) (see e.g. Aubin (1982, §2.1)).

- $B_r(p) \equiv B_r(p; g)$: *open* d_g-ball of radius $r > 0$ and center p.

- $\overline{B}_r(p) \equiv \overline{B}_r(p; g)$: closed d_g-ball of radius $r > 0$ and center p.

- $\mathrm{inj}_g(p)$: injectivity radius of (M, g) at p.

- $\mathrm{inj}_g(M) := \inf\{p \in M : \mathrm{inj}_g(p)\}$.

- μ_g: natural Radon measure on M associated to the Riemannian metric g and a given orientation[7].

The monotonicity formula. We prove a monotonicity formula for Yang–Mills fields due to Price (1983); its proof follows Price's original arguments with almost no modifications.

The metric g enters into the problem as follows. For each fixed point $p \in M$, we let $0 < r_p < \mathrm{inj}_g(p)$ be a small enough radius with the following properties: there are normal coordinates $x^1, \ldots, x^n$ centered at p in the geodesic ball $B_{r_p}(p)$ such that, for some constant $c(p) \geq 0$, the metric components $g_{ij} := g(\partial/\partial x^i, \partial/\partial x^j)$ satisfies the following estimates:

1. $|g_{ij} - \delta_{ij}| \leq c(p)|x|^2$.

2. $|\partial_k g_{ij}| \leq c(p)|x|$.

Note that, from the key properties $g_{ij}(p) = 0$ and $\partial_k g_{ij}(p) = 0$ of normal coordinates, the Taylor expansions of g_{ij} and $\partial_k g_{ij}$ at p show that the constants r_p and $c(p)$ can be chosen depending only on $\mathrm{inj}_g(p)$ and the curvature of g; thus, for instance, when g is flat we can take any $r_p < \mathrm{inj}_g(p)$ and $c(p) = 0$.

It will then be convenient to introduce the following class of Riemannian manifolds:

Definition 3.16 (Bounded Geometry). Let (M^n, g) be a complete Riemannian n-manifold and let $k \in \mathbb{N}_0$. We say that (M^n, g) has **bounded geometry up to order** k when the following conditions are satisfied:

(I) the global injectivity radius of (M^n, g) is positive:

$$\mathrm{inj}_g(M) > 0. \tag{3.17}$$

---

[7]see Example A.11 of Appendix A.

(B_k) the Riemannian curvature R^g and its covariant derivatives up to order k are uniformly bounded: for each $j \in \{0, \dots, k\}$, there exists $c_j \in \mathbb{R}^+$ such that

$$\|(D^g)^j R^g\|_{L^\infty(M)} \leq c_j. \tag{3.18}$$

Example 3.19. (Manifolds of bounded geometry) The following are examples of manifolds with bounded geometry of any (i.e. infinite) order:

- $(\mathbb{R}^n, g_E)$, where g_E is the standard Euclidean metric;

- Any compact Riemannian manifold (M^n, g); indeed, both the injectivity radius and the (derivatives of the) curvature are continuous functions, so these attain maxima and minima on M.

- Riemannian manifolds with a transitive group of isomorphisms (in particular, symmetric spaces). Indeed, the finite injectivity radius and estimates on (derivatives of) the curvature at any single point translate to a uniform estimate for all points under isomorphisms;

- Asymptotically conical (AC) and asymptotically cylindrical (ACyl) Riemannian manifolds with one end. An AC (resp. ACyl) manifold is a non-compact complete Riemannian manifold which outside a compact subset $K \subset M$ is diffeomorphic to a product manifold $(1, \infty)_r \times N$, where (N, g_N) is a closed connected Riemannian manifold, and such that the pullback of g outside K by such diffeomorphism converges (in all derivatives) to the cone metric $dr^2 + r^2 g_N$ at a polynomial rate (resp. cylindrical metric $dr^2 + g_N$ at an exponential rate). For precise definitions, see e.g. Pacini (2013, §6). For the ACyl case the bounded geometry is a simple consequence of the definition together with the fact that a Riemannian product of spaces of bounded geometry is of bounded geometry. As for the AC case, one can appeal to the formulas for the curvature of a warped product metric, cf. Gromoll and Walschap (2009, Proposition 2.2.2).

In this general setup of Riemannian manifolds of bounded geometry, we have the following result on uniform geometric control of small geodesic balls (cf. Eichhorn (1991)):

Theorem 3.20. *Let (M^n, g) be a complete Riemannian n–manifold of bounded geometry up to order $k \geq 1$. Then there exists a constant δ_0 with $0 < \delta_0 < \mathrm{inj}_g(M)$, such that the metric up to its k–th order derivatives and the Christoffel symbols up to its $(k-1)$–th order derivatives are bounded in normal coordinates of radius δ_0 around each $p \in M$, and the bounds are uniform in p.*

In particular, in the base case of manifolds of bounded geometry up to order 1, one can prove the following more precise result (cf. Hebey (2000, p. 16, Theorem 1.3)):

Theorem 3.21. *Let (M^n, g) be a complete Riemannian $n-$manifold of bounded geometry up to order 1, i.e. satisfying (3.17) and (3.18) with $k = 1$. Then there are positive constants $c = c(n, c_0, c_1)$ and $\delta = \delta(n, c_0, c_1)$ depending only on n, c_0 and c_1, such that the components g_{ij} of g in geodesic normal coordinates at p satisfy: for any $i, j, l = 1, \ldots, n$ and any $x \in B_{\delta_0}(0) \subset \mathbb{R}^n$, with $\delta_0 := \min\{\delta, \mathrm{inj}_g(M)\}$,*

(i) $4^{-1}\delta_{ij} \leqslant g_{ij}(\exp_p(x)) \leqslant 4\delta_{ij}$ (as bilinear forms);

(ii) $|g_{ij}(\exp_p(x)) - \delta_{ij}| \leqslant c|x|^2$ and

(iii) $|\partial_l g_{ij}(\exp_p(x))| \leqslant c|x|$.

Moreover, one has that

$$\lim_{(c_0, c_1) \to 0} \delta(n, c_0, c_1) = +\infty \quad and \quad \lim_{(c_0, c_1) \to 0} c(n, c_0, c_1) = 0.$$

Convention 3.22. From now on, whenever (M^n, g) is of bounded geometry up to order 1, we let $0 < \delta_0 < \mathrm{inj}_g(M)$ be given by Theorem 3.21.

Notation 3.23. *In what follows we will always denote by $O(1)$ a quantity bounded by a constant depending only on $n := \dim M$.*

We are now in position to state and prove Tian (2000, Theorem 2.1.2, p. 212):

Theorem 3.24 (Price). *Let $p \in M$, and let r_p and $c(p)$ be as above. Then there exists a nonnegative constant $a \geqslant O(1)c(p)$ such that the following holds. Let $\nabla \in \mathcal{A}(E)$ be a Yang–Mills connection with finite L^2-energy. Then for all $0 < s < r \leqslant r_p$ we have:*

$$e^{ar^2} r^{4-n} \int_{B_r(p)} |F_\nabla|^2 \, dV_g - e^{as^2} s^{4-n} \int_{B_s(p)} |F_\nabla|^2 \, dV_g$$

$$\geqslant 4 \int_{B_r(p) \setminus B_s(p)} e^{a\rho^2} \rho^{4-n} \left| \frac{\partial}{\partial \rho} \lrcorner F_\nabla \right|^2 dV_g.$$

Here $\rho := d_g(p, \cdot)$. Furthermore:

(i) *If $(M, g) = (\mathbb{R}^n, g_0)$, where g_0 denotes the standard flat metric, then we can take $a = 0$ and the above inequality holds for every $p \in M$ and $r \in\,]0, \infty[$.*

(ii) *More generally, if M is of bounded geometry up to order 1, then we can choose a* uniform *constant $a \geqslant 0$ depending only on the geometry of (M, g), so that the above inequality holds for every $p \in M$ and $0 < s < r \leqslant \delta_0$.*

Proof. Without loss of generality we can suppose $r < r_p$; the case $r = r_p$ follows by the obvious approximation argument. Let ξ be a C^∞ cut-off function on the interval $[0, r_p]$, and define the cut-off radial vector field

$$X = X_\xi := \xi(\rho)\rho\frac{\partial}{\partial\rho}.$$

Let $\{e_i\}_{1\leqslant i\leqslant n}$ be an orthonormal local frame of TM near p such that $e_1 = \dfrac{\partial}{\partial\rho}$. Recalling that the unit radial vector field $\dfrac{\partial}{\partial\rho}$ is the velocity of a (radial) geodesic, it follows that

$$D_{\frac{\partial}{\partial\rho}}\frac{\partial}{\partial\rho} = 0.$$

Thus

$$D_{\frac{\partial}{\partial\rho}}X = (\xi\rho)'\frac{\partial}{\partial\rho} = (\xi'\rho + \xi)\frac{\partial}{\partial\rho}.$$

Moreover, for $i \geqslant 2$, we have

$$D_{e_i}X = \xi\rho D_{e_i}\frac{\partial}{\partial\rho} = \xi D_{\rho e_i}\frac{\partial}{\partial\rho} = \xi\sum_{j=2}^{n} b_{ij}e_j,$$

where

$$b_{ij} := g\left(D_{\rho e_i}\frac{\partial}{\partial\rho}, e_j\right) \quad (j = 2,\ldots,n)$$

satisfies

$$|b_{ij} - \delta_{ij}| = O(1)c(p)\rho^2.$$

By straightforward computations, we get

$$|F_\nabla|^2\operatorname{div} X - 4\sum_{i,j=1}^{n}\langle F_\nabla(D_{e_i}X, e_j), F_\nabla(e_i, e_j)\rangle_{\mathfrak{g}} \tag{3.25}$$

$$= \xi'\rho|F_\nabla|^2 + (n-4)\xi|F_\nabla|^2 + O(1)c(p)\rho^2\xi|F_\nabla|^2 - 4\xi'\rho\left|\frac{\partial}{\partial\rho} \lrcorner F_\nabla\right|^2.$$

We choose, for $\tau \in [s, r]$, $\xi(\rho) = \xi_\tau(\rho) = \phi\left(\frac{\rho}{\tau}\right)$ with $\phi = \phi_\varepsilon \in C^\infty([0, \infty[)$, $\varepsilon > 0$ small so that $(1 + \varepsilon)r < r_p$ (recall that $r < r_p$), satisfying: $\phi(t) = 1$ for $t \in [0, 1]$, $\phi(t) = 0$ for $t \in [1 + \varepsilon, \infty[$, and $\phi'(t) \leqslant 0$. Then

$$\tau \frac{\partial}{\partial \tau}(\xi_\tau(\rho)) = -\rho \xi_\tau'(\rho). \tag{3.26}$$

Noting that $\xi_\tau(\rho) \neq 0$ precisely when $\rho \leqslant (1 + \varepsilon)\tau$, it follows from equations (3.25), (3.26) and the variational formula (3.15) that

$$\tau \frac{\partial}{\partial \tau} \int_M \xi_\tau |F_\nabla|^2 \mathrm{d}V_g + \left((4 - n) + O(1)c(p)\tau^2\right) \int_M \xi_\tau |F_\nabla|^2 \mathrm{d}V_g$$
$$= 4\tau \frac{\partial}{\partial \tau} \int_M \xi_\tau \left| \frac{\partial}{\partial \rho} \lrcorner F_\nabla \right|^2 \mathrm{d}V_g.$$

Choosing a nonnegative number $a \geqslant O(1)c(p)$, and multiplying the above equation by $e^{a\tau^2} \tau^{3-n}$, we get

$$\frac{\partial}{\partial \tau} \left(e^{a\tau^2} \tau^{4-n} \int_M \xi_\tau |F_\nabla|^2 \mathrm{d}V_g \right)$$
$$= 4e^{a\tau^2} \tau^{4-n} \frac{\partial}{\partial \tau} \int_M \xi_\tau \left| \frac{\partial}{\partial \rho} \lrcorner F_\nabla \right|^2 \mathrm{d}V_g \tag{3.27}$$
$$+ \left(-O(1)c(p) + 2a\right) e^{a\tau^2} \tau^{5-n} \int_M \xi_\tau |F_\nabla|^2 \mathrm{d}V_g.$$

Since the second term of the RHS of (3.27) is nonnegative (therefore can be dropped), the result follows by integrating over $[s, r]$ and letting $\varepsilon \downarrow 0$.

The final assertions (i) and (ii) follows from Theorem 3.21. $\qquad\square$

Remark 3.28. Following the same arguments of the above proof, Tian (2000, Theorem 2.1.1) proves a slightly generalized version of Theorem 3.24. He needs such version of the formula to perform a proof of the existence of tangent cone measures of blow-up loci Tian (ibid., Lemma 3.2.1). By the direct way we will prove the rectifiability of blow-up loci in Chapter 4, we will not need to provide a separated proof for such existence result; see Theorem 4.14 and Remark 4.15. $\qquad\diamondsuit$

It follows from Price's monotonicity that the map

$$r \mapsto r^{4-n} e^{ar^2} \int_{B_r(p)} |F_\nabla|^2 \mathrm{d}V_g$$

is *non-decreasing* for $r \in \,]0, r_p]$. This will be important in Chapter 4. Moreover, we have the following curious corollary, showing in particular that for

$n \geqslant 5$ every finite-energy Yang–Mills connection over the Euclidean space $\mathbb{R}^n$ is necessarily flat (see Price (1983, Corollary 2, p. 148)).

Corollary 3.29. *Let $\nabla \in \mathcal{A}(E)$ be a Yang–Mills connection with $\mathcal{YM}(\nabla) < \infty$ on a (necessarily trivial) $G-$bundle E over $(\mathbb{R}^n, g_0)$, where g_0 is the standard flat metric. If there is some $x \in \mathbb{R}^n$ such that*[8]

$$\|F_\nabla\|^2_{L^2(B_R(x))} = o(R^{n-4}) \quad \text{as } R \to \infty, \tag{3.30}$$

then ∇ is a flat *connection. In particular, if $n \geqslant 5$ then ∇ is flat.*

Proof. Suppose, by contradiction, that $F_\nabla \neq 0$. Then there exists some $R_0 > 0$ large enough so that

$$\Delta := R_0^{4-n} \|F_\nabla\|^2_{L^2(B_{R_0}(x))} > 0.$$

On the other hand, for each $R \geqslant R_0$, Theorem 3.24 (i) implies that

$$\Delta \leqslant R^{4-n} \|F_\nabla\|^2_{L^2(B_R(x))}.$$

Thus, making $R \to \infty$ and using the hypothesis (3.30) we conclude $\Delta \leqslant 0$ $(\Rightarrow \Leftarrow)$. This proves the main statement.

For the final assertion, simply note that the constant function is $o(R^{n-4})$ when $n \geqslant 5$, and that $\|F_\nabla\|^2_{L^2(B_R(x))} \leqslant \mathcal{YM}(\nabla) = \text{const.} < \infty$ (by hypothesis) for every $x \in \mathbb{R}^n$ and $R > 0$. $\qquad\square$

Remark 3.31. The normalized L^2-norm

$$r^{4-n} \int_{B_r(p)} |F_\nabla|^2 dV_g \tag{3.32}$$

is also known as the *scaling-invariant L^2-norm* of F_∇. Indeed, scale g by λ^2, for some $\lambda \in \mathbb{R}_+$, and let $\widetilde{g} := \lambda^2 g$. It follows easily that $B_{\lambda r}(x; \widetilde{g}) = B_r(x; g)$, for all $x \in M$. Furthermore, the pointwise inner product on $2-$forms scales by λ^{-4}, and the Riemannian volume $n-$forms scales by λ^n; thus,

$$(\lambda r)^{4-n} \int_{B_{\lambda r}(p;\widetilde{g})} |F_\nabla|^2_{\widetilde{g}} dV_{\widetilde{g}} = r^{4-n} \int_{B_r(p)} |F_\nabla|^2 dV_g.$$

◇

---

[8]Here we use the standard little-o notation.

3.3 ε—regularity theorem

Motivated by Schoen's method Schoen (1984, Theorem 2.2) in proving the *a priori* pointwise estimate for stationary harmonic maps, Nakajima (1988, p. 387, Lemma 3.1) combined Price's monotonicity formula together with an appropriate Bochner–Weitzenböck formula to obtain a local L^∞—estimate for Yang–Mills fields satisfying a smallness condition on their normalized L^2—norm over a sufficiently small geodesic ball. Similar results also appears in earlier works by Uhlenbeck, see e.g. Uhlenbeck (1982b, Theorem 3.5) and Uhlenbeck and Yau (1986, Theorem 5.1)[9]. The following statement of Nakajima's result, which we will refer to as the *ε—regularity theorem*, is adapted from Tian (2000, Theorem 2.2.1, p. 213).

Theorem 3.33 (Uhlenbeck–Nakajima). *Let (M^n, g) be a complete oriented Riemannian n—manifold of bounded geometry up to order 1, with dimension $n \geqslant 4$, and let E be a G—bundle over M where G is a compact Lie group. Then there exist scale invariant constants $\varepsilon_0 > 0$ and $C_0 > 0$ such that the following holds. Let $\nabla \in \mathcal{A}(E)$ be a Yang–Mills connection with finite L^2—energy. If $p \in M$ and $0 < r \leqslant \delta_0$ are such that*

$$\varepsilon := r^{4-n} \int_{B_r(p)} |F_\nabla|^2 dV_g < \varepsilon_0,$$

then

$$\sup_{x \in B_{\frac{r}{4}}(p)} |F_\nabla|^2(x) \leqslant C_0 r^{-4} \varepsilon.$$

This theorem is of fundamental importance in compactness theory of Yang–Mills connections in higher dimensions. Following Tian (ibid.), in the next section we will provide a key application of such result (cf. Theorem 3.48) which implies that a sequence of Yang–Mills connections with uniformly L^2—bounded curvatures may fail to have a C^∞_{loc}—convergent subsequence modulo gauge transformations. Indeed, the associated curvatures of such sequences of connections satisfy the hypothesis of the above a priori estimate, provided we look at balls outside a suitable subset $S \subseteq M$ of Hausdorff codimension[10] at least 4, where the curvatures 'blows up'. Only away from S we get uniform local bounds on the curvatures, so that we can apply the standard techniques (cf. section 3.1) to extract a C^∞—convergent subsequence.

The rest of this section is devoted to give a proof of Theorem 3.33. Our proof is based on Tian (ibid., pp. 213-215), which explores the same method

---

[9]The latter refers the reader to a paper by Uhlenbeck (n.d.) that was never published.

[10]See Definition A.14 for the notion of Hausdorff dimension.

of Nakajima's proof Nakajima (1988, Lemma 3.1, pp. 387-388) but fits better in our present notation. We will need the following preliminary lemmas.

Lemma 3.34 (Bochner type estimate). *Suppose (M, g) is an oriented Riemannian n–manifold, and let E is a G–bundle over M. Given $p \in M$ and $0 < r < \mathrm{inj}_g(p)$, there are constants $c', c'' > 0$, where c' depends at most on n and the supremum bound of the Riemannian curvature R^g on $\overline{B}_r(p)$, and c'' depends at most on n and G, such that the following holds. If ∇ is a Yang–Mills connection on E, then*

$$\Delta_g^- |F_\nabla|^2 \geq -c'|F_\nabla|^2 - c''|F_\nabla|^3 \quad on \ B_r(p), \tag{3.35}$$

where $\Delta_g^- := -\mathrm{d}^*\mathrm{d} \colon C^\infty(M) \to C^\infty(M)$ is the Laplace–Beltrami operator with respect to g. In particular, if (M^n, g) is of bounded geometry up to order 1 (thus satisfying (3.18) with $k = 1$), then (3.35) holds on $B_{\delta_0}(p)$ for any $p \in M$, and $c' = c'(n, c_0)$ depends only on n and c_0.

Proof. We start noting that, for any $\xi \in \Omega^k(M, \mathfrak{g}_E)$, we have

$$\begin{aligned}
\Delta_g^- |\xi|^2 &= -\mathrm{d}^*\mathrm{d}\langle \xi, \xi \rangle = -2\mathrm{d}^*\langle \nabla\xi, \xi \rangle \\
&= 2 * \mathrm{d} * \langle \nabla\xi, \xi \rangle = 2 * \mathrm{d}\langle *\nabla\xi, \xi \rangle \\
&= 2 * (\langle \nabla * \nabla\xi, \xi \rangle + \langle *\nabla\xi, \nabla\xi \rangle) = 2 \left(|\nabla\xi|^2 - \langle \nabla^*\nabla\xi, \xi \rangle \right).
\end{aligned}$$

On the other hand, recalling the Bochner–Weitzenböck formula (1.25), for any $\xi \in \Omega^2(M, \mathfrak{g}_E)$ we can write

$$\Delta_\nabla \xi = \nabla^*\nabla\xi + \{R, \xi\} + \{F_\nabla, \xi\},$$

where the brackets $\{,\}$ indicate algebraic multilinear expressions. Combining these facts and using that $\Delta_\nabla F_\nabla = 0$, i.e. ∇ is a Yang–Mills connection, we get:

$$\begin{aligned}
0 = -2\langle \Delta_\nabla F_\nabla, F_\nabla \rangle &= -2\langle \nabla^*\nabla F_\nabla, F_\nabla \rangle + \{R, F_\nabla, F_\nabla\} + \{F_\nabla, F_\nabla, F_\nabla\} \\
&= \Delta_g^- |F_\nabla|^2 - 2|\nabla F_\nabla|^2 + \{R, F_\nabla, F_\nabla\} + \{F_\nabla, F_\nabla, F_\nabla\} \\
&\leq \Delta_g^- |F_\nabla|^2 + \{R, F_\nabla, F_\nabla\} + \{F_\nabla, F_\nabla, F_\nabla\}.
\end{aligned}$$

Therefore

$$\Delta_g^- |F_\nabla|^2 \geq -c'|F_\nabla|^2 - c''|F_\nabla|^3 \quad on \ B_r(p),$$

where $c' > 0$ is a constant depending only on n and the supremum bound of the Riemannian curvature R^g on $\overline{B}_r(p)$, and $c'' > 0$ is a constant depending only on n and G. $\square$

The next lemma, which we state without proof, is a standard mean-value type inequality; see Gilbarg and Trudinger (2001, Theorem 9.20) and Hohloch, Noetzel, and D. A. Salamon (2009, Step 2 in the Proof of Theorem B.1).

Lemma 3.36 (Mean-value inequality). *Suppose (M^n, g) is a complete oriented Riemannian n−manifold of bounded geometry up to order 1 (thus satisfying (3.18) with $k = 1$). Then there exist $c''' > 0$, depending only on n, c_0 and c_1, with the following significance. For every $p \in M$, $r \in (0, \delta_0]$ and smooth function $f : B_r(p) \to [0, \infty)$, one has*

$$\Delta_g^- f \geq -C \quad \Longrightarrow \quad f(p) \leq c''' \left(r^{-n} \int_{B_r(p)} f + C r^2 \right).$$

Now we prove Theorem 3.33. We start noting that since both the normalized L^2−energy of ∇ and the stated L^∞−bound on $|F_\nabla|^2$ are scale invariant (see e.g. Remark 3.31), we can suppose $r = 1$. So we have

$$\varepsilon := \int_{B_1(p)} |F_\nabla|^2 \mathrm{d}V_g < \varepsilon_0, \tag{3.37}$$

and we want to prove that for a sufficiently small $\varepsilon_0 > 0$, depending at most on the geometry of (M^n, g) and G, we get the estimate

$$\sup_{x \in B_{\frac{1}{4}}(p)} |F_\nabla|^2(x) \leq C_0 \varepsilon, \tag{3.38}$$

for some constant $C_0 > 0$ depending at most on the geometry of (M^n, g) and G.

The proof we give here is based on the so-called 'Heinz trick' and follows Walpuski (2017c, Appendix A). Consider the function $\theta \colon \overline{B}_{1/2}(p) \to [0, \infty)$ given by

$$\theta(x) := \left(\frac{1}{2} - d(p, x) \right)^4 |F_\nabla|^2(x).$$

By continuity, θ attains a maximum. Since θ is non-negative and vanishes on the boundary $\partial B_{1/2}(p)$, it achieves its maximum

$$M := \max_{\overline{B}_{1/2}(p)} \theta$$

in the interior of $B_{1/2}(p)$. Now it will be convenient to introduce the following

Notation 3.39. *Henceforth, we write $x \lesssim y$ for $x \leqslant c y$, where $c > 0$ is a generic constant which depends only on the geometry of (M^n, g) and possibly on the structure group G of E.*

We will derive a bound for M of the form $M \lesssim \varepsilon$, from which the assertion of the theorem follows. Let $x_0 \in B_{1/2}(p)$ be a point with $\theta(x_0) = M$, set

$$F_0 := |F_\nabla|^2(x_0)$$

and

$$s_0 := \frac{1}{2}\left(\frac{1}{2} - d(p, x_0)\right).$$

Note that

$$x \in B_{s_0}(x_0) \quad \Rightarrow \quad \left(\frac{1}{2} - d(p, x)\right) \geqslant s_0.$$

Therefore,

$$x \in B_{s_0}(x_0) \quad \Rightarrow \quad |F_\nabla|^2(x) \leqslant s_0^{-4}\theta(x) \leqslant s_0^{-4}\theta(x_0) \lesssim F_0.$$

In particular, it follows from Lemma 3.34 that

$$\Delta_g|F_\nabla|^2 \lesssim |F_\nabla|^2 + |F_\nabla|^3 \lesssim F_0 + F_0^{3/2} \quad \text{on} \quad B_{s_0}(x_0). \tag{3.40}$$

(Here $\Delta_g := -\Delta_g^- = \mathrm{d}^*\mathrm{d}$.) Now it follows from Lemma 3.36 that

$$F_0 \lesssim s^{-n} \int_{B_s(x_0)} |F_\nabla|^2 + s^2(F_0 + F_0^{3/2}), \quad \forall s \leqslant s_0.$$

Hence, the monotonicity (Theorem 3.24) implies

$$F_0 \lesssim s^{-4}\varepsilon + s^2(F_0 + F_0^{3/2}), \quad \forall s \leqslant s_0,$$

which we rewrite as

$$s^4 F_0 \lesssim \varepsilon + s^6(F_0 + F_0^{3/2}), \quad \forall s \leqslant s_0. \tag{3.41}$$

We now have two cases.

$\underline{F_0 \leqslant 1}$: in this case $F_0^{3/2} \leqslant F_0$; hence, for each $s \leqslant s_0$, it follows from (3.41) that $s^4 F_0 \leqslant c\varepsilon + cs^6 F_0$, i.e.

$$s^4 F_0 \leqslant \frac{c\varepsilon}{1 - cs^2}. \tag{3.42}$$

If $cs_0^2 \leqslant 1/2$ then we obtain $s_0^4 F_0 \lesssim \varepsilon$, so that

$$M = \theta(x_0) \lesssim s_0^4 F_0 \lesssim \varepsilon;$$

otherwise, setting $s := (2c)^{-1/2} \leqslant s_0$ and plugging into (3.42) yields $F_0 \lesssim \varepsilon$, and thus $M \lesssim s_0^4 F_0 \lesssim \varepsilon$, since $s_0 \leqslant 1$.

$\underline{F_0 > 1}$: in this case $F_0 \leqslant F_0^{3/2}$, so that from (3.41) we derive

$$s^4 F_0 \leqslant c\varepsilon + cs^6 F_0^{3/2}, \quad \forall s \leqslant s_0. \tag{3.43}$$

Thus, setting $t = t(s) := sF_0^{1/4}$, the inequality (3.43) can be expressed as

$$t^4(1 - ct^2) \leqslant c\varepsilon.$$

Now we can choose $\varepsilon_0 > 0$ sufficiently small, relatively to an amount depending only on c, hence only on the geometry of (M, g) and G, so that, for $\varepsilon \leqslant \varepsilon_0$, the corresponding equation $t^4(1-ct^2) = c\varepsilon$ has four small (real) roots $t_1, \ldots, t_4$, which are approximately $\pm(c\varepsilon)^{1/4}$, and two large (complex) roots. Since $t(0) = 0$ and t is continuous, $t(s)$ must be less than the smallest positive (real) root for each $s \in [0, s_0]$. Therefore, $t(s) \lesssim \varepsilon^{1/4}$ for all $s \in [0, s_0]$; in particular, $M \lesssim \varepsilon$. This completes the proof of the theorem.

3.4 Convergence away from the blow-up locus

Let (M^n, g) be a complete oriented Riemannian $n-$manifold of bounded geometry up to order 1 and dimension $n \geqslant 4$, and let E be a $G-$bundle over M where G is a compact Lie group. As we have seen in the last section, the $\varepsilon-$regularity theorem (Theorem 3.33) provides *a priori* local $L^\infty-$bounds on the curvature of a Yang–Mills connection ∇ on E provided its normalized L^2-energy is sufficiently small. Given a sequence $\{\nabla_i\}$ of Yang–Mills connections with uniformly L^2-bounded curvatures, this previous result allows us to define a closed subset of points in M around which the C^1-convergence (modulo gauge) of any subsequence necessarily has to fail and outside of which there is a subsequence converging (modulo gauge) in C_{loc}^∞. (cf. Nakajima (1988) and Tian (2000, Lemma 3.1.3)).

Definition 3.44. The **blow-up locus** (or **energy concentration set**[11]) $S = S(\{\nabla_i\})$ of $\{\nabla_i\}$ is the set

$$S := \bigcap_{0 < r \leqslant \delta_0} \left\{ x \in M : \liminf_{i \to \infty} e^{ar^2} r^{4-n} \int_{B_r(x)} |F_{\nabla_i}|^2 \mathrm{d}V_g \geqslant \varepsilon_0 \right\}, \tag{3.45}$$

---

[11] This terminology is used in the context of harmonic map theory, see e.g. Lin (1999, p. 787).

where $a \geqslant 0$ is the constant given by the monotonicity formula (Theorem 3.24 (ii)) and $\varepsilon_0 > 0$ is the constant given by the ε–regularity theorem (Theorem 3.33).

Remark 3.46. We caution the reader that the notation and terminology used here differ from those used in the main reference Tian (2000). Indeed, Tian denotes the set $S(\{\nabla_i\})$ by $S_b(\{\nabla_i\})$ and reserves the name 'blow-up locus' for a certain subset of $S_b(\{\nabla_i\})$ which he denotes by S_b. The latter, in turn, is what we will define to be the 'bubbling locus' Γ of $\{\nabla_i\}$ (see Definition 4.9 and Remark 4.12). In fact, we follow the terminology and notations used in a more recent work of Walpuski (2017c), which makes the same sort of blow-up analysis explored in Tian's paper, based on Lin's paper Lin (1999), albeit in the context of Fueter sections. The reader will find out in Chapter 4 the main reason for the terminology 'bubbling locus' (cf. Theorem 4.31 and Proposition 4.37). $\diamondsuit$

Notation 3.47. *Henceforth, we denote by $\mathcal{H}^{n-4}$ the $(n-4)$–dimensional Hausdorff measure*[12] *of the connected Riemannian n–manifold (M, g).*

Theorem 3.48 (Uhlenbeck–Nakajima). *Let (M^n, g) be a complete oriented Riemannian manifold of bounded geometry up to order 1 and dimension $n \geqslant 4$, and let $E \to M$ be a G–bundle with compact structure group. Suppose $\{\nabla_i\} \subseteq \mathcal{A}(E)$ is a sequence of Yang–Mills connections with uniformly L^2–bounded curvatures, say $\mathcal{YM}(\nabla_i) \leqslant \Lambda$. Then:*

(i) *The blow-up locus S of $\{\nabla_i\}$ (cf. Definition 3.44) is a closed subset of M and $\mathcal{H}^{n-4}(S) \leqslant C(n, g, G, \Lambda) < \infty$.*

(ii) *There exist a subsequence of $\{\nabla_i\}$, still denoted by $\{\nabla_i\}$, a sequence of gauge transformations $g_i \in \mathcal{G}(E|_{M \setminus S})$, and a smooth Yang–Mills connection ∇ on the restriction $E|_{M \setminus S}$, such that $g_i^* \nabla_i$ converges to ∇ in C_{loc}^∞–topology outside S.*

(iii) *$M \setminus S$ is the maximal open subset on which a subsequence $\{\nabla_i\}_{i \in I \subset \mathbb{N}}$ can converge in C_{loc}^∞.*

Proof. (i): We divide the proof into two parts:

(a) S is closed.

(b) $\mathcal{H}^{n-4}(S) \leqslant C(n, g, G, \Lambda) < \infty$.

---

[12]See Definition A.10 and Example A.11 of Appendix A

To prove (a), pick $x_0 \in M \setminus S$; there exist $0 < r_0 \leqslant \delta_0$ and a sequence $i_j \to \infty$ such that

$$\sup_j r_0^{4-n} \int_{B_{r_0}(x_0)} |F_{\nabla_{i_j}}|^2 dV_g < \varepsilon_0.$$

Applying the curvature estimate (3.33), we get

$$\sup_j \sup_{x \in B_{\frac{r_0}{4}}(x_0)} |F_{\nabla_{i_j}}(x)|^2 < C_0 \varepsilon_0 r_0^{-4}.$$

Thus, if we let $K > 0$ be such that $\mu_g(B_r(x_0)) \leqslant r^n K$ for all $r \leqslant r_0/8$, we have

$$\sup_j \sup_{x \in B_{\frac{r_0}{8}}(x_0)} e^{ar^2} r^{4-n} \int_{B_r(x)} |F_{\nabla_{i_j}}|^2 dV_g \leqslant e^{ar^2} r^4 K C_0 \varepsilon_0 r_0^{-4}$$

$$= \mathrm{const.} e^{ar^2} r^4.$$

In particular, there exists some $0 < r \leqslant r_0/8$ small enough that

$$\sup_j \sup_{x \in B_{\frac{r_0}{8}}(x_0)} e^{ar^2} r^{4-n} \int_{B_r(x)} |F_{\nabla_{i_j}}|^2 dV_g < \frac{\varepsilon_0}{2},$$

whence we conclude that $B_{\frac{r_0}{8}}(x_0) \subseteq M \setminus S$.

Now we prove (b). Let $0 < \delta < \min\{1, \delta_0\}$ be arbitrary. Then we can find a countable covering $\{B_{5r_\alpha}(x_\alpha)\}$ of S such that $x_\alpha \in S$ and $10 r_\alpha \leqslant \delta$ for each α, and (†) the $B_{r_\alpha}(x_\alpha)$ are pairwise disjoint (see Theorem A.16). Thus

$$\sum_\alpha r_\alpha^{n-4} \leqslant \frac{e^{ar_\alpha^2}}{\varepsilon_0} \sum_\alpha \liminf_{i\to\infty} \int_{B_{r_\alpha}(x_\alpha)} |F_{\nabla_i}|^2 dV_g \quad (x_\alpha \in S \text{ and } r_\alpha \leqslant \delta_0)$$

$$\leqslant \frac{e^{ar_\alpha^2}}{\varepsilon_0} \liminf_{i\to\infty} \sum_\alpha \int_{B_{r_\alpha}(x_\alpha)} |F_{\nabla_i}|^2 dV_g$$

$$\leqslant \frac{e^{ar_\alpha^2}}{\varepsilon_0} \liminf_{i\to\infty} \int_{\bigcup_\alpha B_{r_\alpha}(x_\alpha)} |F_{\nabla_i}|^2 dV_g \quad (\dagger)$$

$$\leqslant \frac{e^a}{\varepsilon_0} \liminf_{i\to\infty} \int_M |F_{\nabla_i}|^2 dV_g \quad (r_\alpha \leqslant \delta \leqslant 1)$$

$$\leqslant \frac{e^a \Lambda}{\varepsilon_0}. \quad (\mathcal{YM}(\nabla_i) \leqslant \Lambda \text{ for all } i)$$

Since $\{B_{5r_\alpha}(x_\alpha)\}$ covers S and $10r_\alpha \leqslant \delta$, we get

$$\mathcal{H}_\delta^{n-4}(S) \leqslant \sum_\alpha 5^{n-4} r_\alpha^{n-4} \leqslant \frac{5^{n-4} e^a \Lambda}{\varepsilon_0} =: C,$$

where $C = C(n, g, G, \Lambda) > 0$ is *independent* of δ. Thus, it follows that

$$\mathcal{H}^{n-4}(S) = \lim_{\delta \downarrow 0} \mathcal{H}_\delta^{n-4}(S) \leqslant C.$$

(ii): By the proof of (i)-(a) above, for each $x \in M \setminus S$ there exist a neighborhood U_x of x in $M \setminus S$ and a subsequence $\{i_j^{(x)}\} \subseteq \{i\}$ such that $|F_{\nabla_{i_j^{(x)}}}|$ is uniformly bounded on U_x. Thus, invoking Theorem 3.10, we can find a single subsequence $\{i_j\} \subseteq \{i\}$, gauge transformations g_{i_j} of E over $M \setminus S$ and a smooth Yang–Mills connection ∇ on $E|_{M \setminus S}$ such that $g_{i_j}^* \nabla_{i_j}$ converges to ∇ in C_{loc}^∞–topology outside S.

(iii): We prove that if S' is a closed subset of M such that a subsequence $\{\nabla_i\}_{i \in I \subset \mathbb{N}}$ converges in C_{loc}^∞ outside S', then $S' \subset S$. Indeed, suppose that $\{\nabla_i\}_{i \in I \subset \mathbb{N}}$ converges in C^1 in a neighborhood of $x \in M$. Then $|F_{\nabla_i}|$ is uniformly bounded in this neighborhood. Hence, by the same reasoning of the proof of (i)-(a), there is a slightly smaller neighborhood of x which is contained in $X \setminus S$; in particular, $x \in X \setminus S$. $\qquad\square$

Remark 3.49 (Uhlenbeck compactness of the moduli space of flat connections). An easy application of the above theorem shows that the moduli space of flat connections $\mathcal{M}_{\mathrm{flat}}(E) := \{\nabla \in \mathcal{A}(E) : F_\nabla = 0\}/\mathcal{G}(E)$ is compact in the natural topology of C_{loc}^∞–convergence modulo gauge transformations. Indeed, if $\{\nabla_i\}$ is a sequence of flat connections on the G–bundle E, then $\{\nabla_i\}$ trivially satisfies the hypothesis of Theorem 3.48 and, furthermore, $S(\{\nabla_i\}) = \emptyset$ (cf. Definition 3.45). Thus, after passing to a subsequence, we can find a Yang–Mills connection $\nabla \in \mathcal{A}(E)$ and gauge transformations $g_i \in \mathcal{G}(E)$ such that $g_i^* \nabla_i$ converges to ∇ in C_{loc}^∞–topology on M. Clearly, the limit connection ∇ is necessarily flat, thereby proving the claim. $\qquad\Diamond$

Remark 3.50. A special case is when $n = 4$: here $\mathcal{H}^{n-4} = \mathcal{H}^0$ is simply the counting measure, hence Theorem 3.48 (i) implies that the blow-up set of any sequence of Yang–Mills connections with uniformly bounded L^2–energy on a G–bundle over a complete, oriented Riemannian 4–manifold of bounded geometry is necessarily *finite*. $\qquad\Diamond$

In the next chapter, we will examine the causes of this noncompactness phenomenon along the blow-up set.

$$\Huge 4 \qquad\qquad \textit{Structure of}$$
$$\textit{blow-up loci}$$

In this chapter we will make use of some basic results in geometric measure theory. For the sake of completeness and textual linearity, we collect these results together with the relevant definitions in Appendix A.

Let us study the structure of the blow-up set S of a sequence of Yang–Mills connections $\{\nabla_i\}$ with uniformly bounded energy. More specifically, we examine the causes of the formation of S, its rectifiability and some of its geometry. We follow mainly chapters 3 and 4 of Tian's paper Tian (2000), with some adaptations guided by the blow-up analysis approach in Walpuski (2017c).

In Section 4.1, we start noting that, after passing to a subsequence if necessary, the Radon measures $\mu_i := |F_{\nabla_i}|^2 dV_g$ have a weak* limit $\mu = |F_\nabla|^2 dV_g + v$, where ∇ is the C^∞_{loc}–limit of ∇_i outside S (after passing to a subsequence and modulo gauge) and $v \ll \mathcal{H}^{n-4} \lfloor S$ is some (nonnegative) Radon measure singular with respect to μ_g. Moreover, the $(n-4)$–dimensional density function Θ of μ exists, is upper semi-continuous, vanishes outside S, is bounded and $\Theta(x) \geqslant \varepsilon_0$ for all $x \in S$, where $\varepsilon_0 > 0$ is given by Theorem 3.33. We then proceed to show that S decomposes into two closed pieces:

$$S = \Gamma \cup \mathrm{sing}(\nabla),$$

where $\Gamma := \mathrm{supp}(v)$ and $\mathrm{sing}(\nabla)$ is the support of the $(n-4)$–dimensional upper density of $|F_\nabla|^2 dV_g$. Further, $\mathrm{sing}(\nabla)$ is shown to be an $\mathcal{H}^{n-4}$–negligible set. Next, in Section 4.2, we show a first regularity result for the blow-up locus:

Γ is a countably $\mathcal{H}^{n-4}$*—rectifiable* set, i.e. at $\mathcal{H}^{n-4}$—a.e. $x \in \Gamma$ the approximate $(n-4)$—dimensional tangent space $T_x\Gamma$ exists, and ν can be written as $\nu = \Theta(\mu,\cdot)\mathcal{H}^{n-4}\lfloor\Gamma$.

Section 4.3 is the core of this chapter. We analyze the behavior of ∇_i for $i \gg 1$ near a smooth point $x \in \Gamma$, i.e. a point $x \notin \mathrm{sing}(\nabla)$ at which $T_x\Gamma$ is well-defined. The main result is that, at any such x, there is a blowing up of the sequence $\{\nabla_i\}$ around the point x whose limit $B(x)$ is a non-flat Yang–Mills connection on $T_x M$ which is, modulo gauge, the pull-back of a connection $I(x)$ on $T_x\Gamma^\perp$ satisfying the energy inequality

$$\mathcal{YM}(I(x)) \leqslant \Theta(\mu, x).$$

At this stage, we know that at each point x of the blow-up locus S the sequence $\{\nabla_i\}$ loses energy via bubbling and/or develops a singularity.

In Section 4.4 we turn to the case in which $\{\nabla_i\}$ is a sequence of $\varXi$—antiselfdual instantons, for some smooth $(n-4)$—form $\varXi$ assumed to be a calibration on the base manifold. In this case, we are able to show that, at any smooth point $x \in \Gamma$, the tangent space $T_x\Gamma$ is calibrated by $\varXi$, and each bubble $I(x)$ is a non-flat ASD instanton. Moreover, when $G = \mathrm{SU}(r)$, we prove that the natural $(n-4)$—current $C(\Gamma, \Theta)$ defined by the triple $(\Gamma, \varXi, \frac{1}{8\pi^2}\Theta)$ satisfy the following conservation of the instanton charge density:

$$\mathrm{tr}(F_{\nabla_i} \wedge F_{\nabla_i}) \rightharpoonup \mathrm{tr}(F_\nabla \wedge F_\nabla) + 8\pi^2 C(\Gamma, \Theta).$$

These results, due to G. Tian (2000), show a striking relationship between gauge theory and calibrated geometry: if $\varXi$ is a calibration then the blow-up locus of a sequence of $\varXi$—anti-selfdual instantons defines a $\varXi$—calibrated integer rectifiable current, i.e. a generalized (possibly very singular) $\varXi$—calibrated submanifold.

Convention 4.1. Throughout this chapter, unless otherwise stated, (M, g) denotes a connected, complete, oriented Riemannian n—manifold of bounded geometry up to order 1, with $n \geqslant 4$, and E denotes a G—bundle over M, where G is a compact Lie group.

4.1 Decomposition of blow-up loci

From now on we consider a sequence of Yang–Mills connections $\{\nabla_i\}$ such that

$$\mathcal{YM}(\nabla_i) \leqslant \Lambda, \tag{4.2}$$

for some uniform constant $\Lambda > 0$. By Uhlenbeck–Nakajima's Theorem 3.48, we may assume, after passing to a subsequence if necessary, that the ∇_i converges (modulo gauge) to a Yang–Mills connection ∇ in C^∞_{loc} outside the blow-up locus $S = S(\{\nabla_i\})$ (cf. Definition 3.44). In this section, we shall investigate the causes of the formation of the set S.

A key idea to study S is to consider the sequence of Radon measures $\{\mu_i\}$ defined by

$$\mu_i := |F_{\nabla_i}|^2 \mu_g.$$

Indeed, note that we can write S as

$$S = \bigcap_{0 < r \leq \delta_0} \left\{ x \in M : \liminf_{i \to \infty} e^{ar^2} r^{4-n} \mu_i(B_r(x)) \geq \varepsilon_0 \right\}.$$

By the uniform L^2–energy bound hypothesis (4.2), the sequence $\{\mu_i\}$ is of bounded mass; therefore, it converges weakly* to a Radon measure μ on M (cf. Theorem A.28). Then, by Fatou's lemma we get

$$\int_M f|F_\nabla|^2 \mathrm{d}V_g \leq \liminf_{i \to \infty} \int_M f|F_{\nabla_i}|^2 \mathrm{d}V_g = \int_M f \, \mathrm{d}\mu,$$

for all $f \in C^0_c(M)$. Thus, applying Riesz's representation theorem (see Remark A.26), there exists a unique (nonnegative) Radon measure ν on M such that[1]

$$\mu = |F_\nabla|^2 \mu_g + \nu.$$

ν is called the **defect measure**.

Lemma 4.3. $\nu(M \setminus S) = 0$. *In particular,* $\mathrm{supp}(\nu) \subseteq S$ *and* ν *is singular with respect to* μ_g.

Proof. Since $M \setminus S$ is an open set, by Theorem A.24 it suffices to prove that for every $f \in C^0_c(X)$ such that $\mathrm{supp}(f) \subseteq M \setminus S$ and $\|f\|_\infty \leq 1$, we have

$$\lim_{i \to \infty} \int_M f|F_{\nabla_i}|^2 \mathrm{d}V_g = \int_M f|F_\nabla|^2 \mathrm{d}V_g.$$

Denote by K the (compact) support of f in $M \setminus S$, and consider, for each $i \in \mathbb{N}$, the functions

$$\phi_i := f|F_{\nabla_i}|^2 \quad \text{and} \quad \psi_i := \chi_K |F_{\nabla_i}|^2.$$

---

[1] More explicitly, ν is the unique Radon measure on M such that

$$\int_M f \, \mathrm{d}\nu = \int_M f \, \mathrm{d}\mu - \int_M f|F_\nabla|^2 \mathrm{d}V_g, \quad \forall f \in C^0_c(M).$$

It's clear that $|\phi_i| \leq \psi_i$ for each $i \in \mathbb{N}$.

By hypothesis, there exists a sequence $\{g_i\} \subseteq \mathcal{G}(E|_{M \setminus S})$ such that $g_i^* \nabla_i \to \nabla$ in C^∞_{loc} on $M \setminus S$. Thus, using the invariance $|F_{\nabla_i}| = |F_{g_i^* \nabla_i}|$, (and the fact that $\mu_g(S) = 0$ — see below,) when $i \to \infty$ we have

$$\phi_i \to \phi := f|F_\nabla|^2 \quad \mu_g\text{--a.e. on } M,$$

and

$$\psi_i \to \psi := \chi_K |F_\nabla|^2 \quad \text{uniformly on } M.$$

Moreover, from the uniform bound $\mathcal{YM}(\nabla_i) \leq \Lambda$, we automatically have $\phi_i, \phi, \psi_i, \psi \in L^1(\mu_g)$.

Finally, note that the uniform convergence $\psi_i \to \psi$ implies

$$\lim_{i \to \infty} \int_M \psi_i \, dV_g = \int_M \psi \, dV_g,$$

since the ψ_i are supported in a compact set and μ_g is Radon. Therefore, from a well-known version of the dominated convergence theorem (see Folland (2013, p. 59, exercise 20.)), it follows that

$$\lim_{i \to \infty} \int_M \phi_i \, dV_g = \int_M \phi \, dV_g,$$

as we wanted.

The assertion that $\mathrm{supp}(\nu) \subseteq S$ follows from the fact that $M \setminus S$ is an *open* subset with $\nu(M \setminus S) = 0$ (cf. Definition A.2). Moreover, the fact that S is closed and has finite $(n - 4)$–dimensional Hausdorff measure (cf. Theorem 3.48) implies that $\mu_g(S) = \mathcal{H}^n(S) = 0$, so that ν is singular with respect to μ_g. $\qquad\square$

In what follows we will see that the weak* limit measure μ, and its components $|F_\nabla|^2 \mu_g$ and ν, play a fundamental role in the study of S. The next lemma states some crucial facts about μ (compare with Tian (2000, (proof of) Lemma 3.1.4, pp. 221-223)).

Lemma 4.4. *The measure μ and its density function $\Theta(\mu, \cdot)$ have the following properties:*

(i) μ inherits the monotonicity property: for all $x \in M$,

$$0 < s < r \leq \delta_0 \quad \Rightarrow \quad e^{as^2} s^{4-n} \mu(B_s(x)) \leq e^{ar^2} r^{4-n} \mu(B_r(x)),$$

where $a \geqslant 0$ is the constant given in Theorem 3.24. In particular, the $(n-4)$-density of μ at x,

$$\Theta(\mu, x) := \Theta^{n-4}(\mu, x) = \lim_{r \downarrow 0} r^{4-n} \mu(B_r(x)),$$

exists and is bounded by $e^{a \delta_0^2} \delta_0^{n-4} \Lambda$ for every $x \in M$.

(ii) $\Theta(\mu, \cdot)$ defines an upper semi-continuous function on M which vanishes outside S and satisfies $\Theta(x) \geqslant \varepsilon_0$ for all $x \in S$.

Proof.

(i) Let $x \in M$ and $0 < s < r \leqslant \delta_0$. Then, for each $i \in \mathbb{N}$, we know from Price's monotonicity (Theorem 3.24) that

$$e^{as^2} s^{4-n} \mu_i(B_s(x)) \leqslant e^{ar^2} r^{4-n} \mu_i(B_r(x)). \tag{4.5}$$

Also, since $\mu_i \rightharpoonup \mu$, we have (cf. Theorem A.29 (i)):

$$\mu(B_s(x)) \leqslant \liminf_{i \to \infty} \mu_i(B_s(x)). \tag{4.6}$$

Now let $\mathscr{R}_{x,\delta_0}(\mu) \subseteq \,]0, \delta_0]$ be defined as in Theorem A.29 (iv). If $r \notin \mathscr{R}_{x,\delta_0}(\mu)$, then (4.5) and (4.6) immediately imply

$$e^{as^2} s^{4-n} \mu(B_s(x)) \leqslant e^{ar^2} r^{4-n} \mu(B_r(x)).$$

The general case follows by an approximation argument. Indeed, since $\mathscr{R}_{x,\delta_0}(\mu)$ is countable, if $r \in \mathscr{R}_{x,\delta_0}(\mu)$ then we can find $\{r_j\} \subseteq \,]s, r]$, with $r_j \uparrow r$, such that $r_j \notin \mathscr{R}_{x,\delta_0}(\mu)$ for all $j \in \mathbb{N}$. Therefore, on the one hand, by the monotone convergence theorem,

$$\mu(B_r(x)) = \lim_{j \to \infty} \mu(B_{r_j}(x)). \tag{4.7}$$

On the other hand, since $r_j \notin \mathscr{R}_{x,\delta_0}(\mu)$ for all $j \in \mathbb{N}$,

$$e^{as^2} s^{4-n} \mu(B_s(x)) \leqslant e^{ar_j^2} r_j^{4-n} \mu(B_{r_j}(x)), \quad \forall j \in \mathbb{N}.$$

Now make $j \to \infty$ in the above inequality and use (4.7). This completes the proof of the monotonicity property.

From the above it is immediate that

$$0 \leqslant \Theta_*(\mu, x) = \Theta^*(\mu, x) \leqslant e^{a \delta_0^2} \delta_0^{4-n} \Lambda, \quad \forall x \in M.$$

This completes the proof of (i).

(ii) To see that $\Theta(\mu, \cdot)$ is upper semi-continuous, suppose $\{x_m\}$ is a sequence of points in M with $x_m \to x \in M$ as $m \to \infty$. Let $\varepsilon > 0$ and $0 < r \leq \delta_0$. Then, by the monotonicity (i), for $m \gg 1$ we have

$$\Theta(\mu, x_m) \leq e^{ar^2} r^{4-n} \mu(B_r(x_m)) \leq e^{ar^2} r^{4-n} \mu(B_{r+\varepsilon}(x)).$$

Hence, $\limsup_{m \to \infty} \Theta(\mu, x_m) \leq e^{ar^2} r^{4-n} \mu(B_r(x))$. Taking the limit as $r \downarrow 0$, we arrive at the desired conclusion.

Now let $x \in S$. We shall prove that $\Theta(\mu, x) \geq \varepsilon_0$; in fact, we claim that

$$e^{ar^2} r^{4-n} \mu(B_r(x)) \geq \varepsilon_0, \quad \forall r \in \,]0, \delta_0]. \tag{4.8}$$

Indeed, if $r \notin \mathscr{R}_{x,\delta_0}(\mu)$ then (cf. Theorem A.29 (iv))

$$e^{ar^2} r^{4-n} \mu(B_r(x)) = \lim_{i \to \infty} e^{ar^2} r^{4-n} \mu_i(B_r(x)),$$

so that the inequality (4.8) holds due to $x \in S$. If $r \in \mathscr{R}_{x,\delta_0}(\mu)$, we can proceed by an approximation argument just as in the proof of (i): since $\mathscr{R}_{x,\delta_0}(\mu)$ is countable, we can find $\{r_j\} \subseteq \,]0, r[$, with $r_j \uparrow r$, such that $r_j \notin \mathscr{R}_{x,\delta_0}(\mu)$ for all $j \in \mathbb{N}$. Then, on the one hand, by the monotone convergence theorem,

$$\mu(B_r(x)) = \lim_{j \to \infty} \mu(B_{r_j}(x)).$$

On the other hand, by the choice of the sequence $\{r_j\}$ and the first part,

$$e^{ar_j^2} r_j^{4-n} \mu(B_{r_j}(x)) \geq \varepsilon_0, \quad \forall j \in \mathbb{N}.$$

Thus, letting $j \to \infty$ proves (4.8) as we wanted.

Finally, let $x \notin S$. We want to show that $\Theta(\mu, x) = 0$. By definition of S, there exists $r_0 \in (0, \delta_0]$ and a subsequence $i_j \to \infty$ such that

$$e^{ar_0^2} r_0^{4-n} \mu_{i_j}(B_{r_0}(x)) < \varepsilon_0.$$

By the ε–regularity theorem it follows that

$$\sup_j \sup_{y \in B_{\frac{r_0}{4}}(x)} |F_{\nabla_{i_j}}|^2(y) \leq C \varepsilon_0 r_0^{-4}.$$

Therefore, for every $r \in \,]0, r_0/4[$ one has

$$r^{4-n} \mu_{i_j}(B_r(x)) \lesssim r_0^{-4} r^4.$$

Thus

$$\Theta(\mu, x) = \lim_{r \downarrow 0} r^{4-n} \mu(B_r(x))$$

$$\leqslant \lim_{r \downarrow 0} \liminf_{i \to \infty} r^{4-n} \mu_i(B_r(x))$$

$$\lesssim \lim_{r \downarrow 0} r_0^{-4} r^4 = 0,$$

as we wanted. This concludes the proof of (ii).

$\square$

With the above results in mind, we now introduce some terminology.

Definition 4.9. Let $\{\nabla_i\}$, ∇, $\{\mu_i\}$ and $\mu = |F_\nabla|^2 \mu_g + \nu$ be as above. Then

$$\Gamma := \mathrm{supp}(\nu)$$

is called the **bubbling locus**, and $\Theta(\mu, \cdot)$ is its **multiplicity**. We call

$$\mathrm{sing}(\nabla) := \left\{ x \in M : \Theta^*(\nabla, x) := \limsup_{r \downarrow 0} r^{4-n} \int_{B_r(x)} |F_\nabla|^2 dV_g > 0 \right\}$$

the **singular set** of ∇.

Proposition 4.10. $\mathcal{H}^{n-4}(\mathrm{sing}(\nabla)) = 0$ *for* $\mathcal{H}^{n-4}-$*a.e.* $x \in M$.

Proof. Given $\varepsilon > 0$, we set

$$E_\varepsilon := \left\{ x \in M : \Theta^*(\nabla, x) > \varepsilon \right\}.$$

By arbitrariness of $\varepsilon > 0$, it suffices to show that $\mathcal{H}^{n-4}(E_\varepsilon) = 0$. Given $\delta > 0$, we can find a countable covering of E_ε by balls $B_{5r_\alpha}(x_\alpha)$ with centers $x_\alpha \in E_\varepsilon$ and radius $5r_\alpha \leqslant \delta$, such that the balls $B_{r_\alpha}(x_\alpha)$ are pairwise disjoint. Moreover, we can arrange that

$$r_\alpha^{4-n} \int_{B_{r_\alpha}(x_\alpha)} |F_\nabla|^2 dV_g > \varepsilon.$$

Since ∇ is smooth on $M \setminus S$, we must have $E_\varepsilon \subset S$. Thus,

$$\sum_\alpha r_\alpha^{n-4} \leqslant \frac{1}{\varepsilon} \sum_\alpha \int_{B_{r_\alpha}(x_\alpha)} |F_\nabla|^2 dV_g \leqslant \int_{N_\delta(S)} |F_\nabla|^2 dV_g,$$

where $N_\delta(S) := \{x \in M : d_g(x, S) < \delta\}$. Since $\mathcal{H}^n(S) = 0$, the right-hand side goes to 0 as $\delta \to 0$. Therefore $\mathcal{H}^{n-4}(E_\varepsilon) = 0$, completing the proof.

$\square$

Proposition 4.11 (Decomposition of the blow-up locus). *The blow-up locus S decomposes as*

$$S = \Gamma \cup \mathrm{sing}(\nabla).$$

Proof.

- ($\supseteq$) : By Lemma 4.3, we have $\Gamma \subseteq S$. So, it suffices to prove that $\mathrm{sing}(\nabla) \subseteq S$. Now, clearly if ∇ is smooth in a neighborhood of x, then $x \notin \mathrm{sing}(\nabla)$. Since S is closed and ∇ is smooth on $M \setminus S$, the desired inclusion follows.

- ($\subseteq$) : Let $x \in S$. Then, by Lemma 4.4 (ii), we know that $\Theta(\mu, x) \geq \varepsilon_0 > 0$. Since $\mu = |F_\nabla|^2 \mu_g + \nu$, we have:

  - if $x \notin \Gamma$, then $\Theta(\nu, x) = 0$ (cf. Remark A.20) and, therefore, $\Theta^*(\nabla, x) = \Theta(\mu, x) \geq \varepsilon_0 > 0$; thus $x \in \mathrm{sing}(\nabla)$.
  - if $x \notin \mathrm{sing}(\nabla)$, i.e. if $\Theta^*(\nabla, x) = 0$, then $\Theta(\nu, x) = \Theta(\mu, x) \geq \varepsilon_0 > 0$, so that $x \in \Gamma$ (again by Remark A.20).

$\square$

Notice that, by the gauge invariance of $|F_\nabla|$, the singular set $\mathrm{sing}(\nabla)$ is invariant under gauge transformations, so that it consists of *non-removable* singularities of ∇. On the other hand, since we can write the energy concentration set as $S = \{x \in M : \Theta(\mu, x) \geq \varepsilon_0\}$ (cf. Lemma 4.4 (ii)), one should interpret $\Theta(\nu, x)$ as the energy density lost by the sequence $\{\nabla_i\}$ around $x \in S$. Thus, the above result shows that the noncompactness along S has two sources: one involving loss of energy and one involving the formation of non-removable singularities.

Remark 4.12. As a consequence of Proposition 4.11, we can now establish the equality between Tian's definition of the blow-up locus (cf. Tian (2000, (3.1.11), p. 223)) and our definition of the bubbling locus (cf. Definition 4.9), as claimed in Remark 3.46. Define[2]

$$S_b := \overline{\{x \in S : \Theta^*(\nabla, x) = 0\}}.$$

We want to show that

$$\Gamma = S_b.$$

Since S is closed and $\mathrm{sing}(\nabla) \subseteq S$, it immediately follows from the definitions of S_b and $\mathrm{sing}(\nabla)$ that

$$S = S_b \cup \mathrm{sing}(\nabla).$$

---

[2]Since $x \in S$ implies $\Theta(\mu, x) > 0$ (Lemma 4.4 (ii)), this is precisely Tian's original definition of S_b.

Moreover, using the characterization of Lemma 4.4 (ii) for S and that $\mu = |F_\nabla|^2 dV_g + \nu$, we have:

$$S_b = \overline{\{x \in M : 0 < \Theta(\mu, x) = \Theta(\nu, x)\}}.$$

Recalling Remark A.20 and the fact that $\Gamma = \mathrm{supp}(\nu)$ is closed, it follows that $S_b \subseteq \Gamma$. Now, by Proposition 4.11, we know that $S = \Gamma \cup \mathrm{sing}(\nabla)$. So it remains to show that $\mathrm{sing}(\nabla) \setminus S_b \subseteq \mathrm{sing}(\nabla) \setminus \Gamma$. Let $x \in \mathrm{sing}(\nabla) \setminus S_b$. Then, there exists an open subset U of M such that $x \in U \cap S \subseteq \mathrm{sing}(\nabla)$. Thus

$$\left[\mathcal{H}^{n-4} \lfloor S\right](U) \leqslant \mathcal{H}^{n-4}(\mathrm{sing}(\nabla)) = 0,$$

where in the last step we used Proposition 4.11. Now since $\Theta^*(\nu, \cdot) \leqslant \Theta(\mu, \cdot) < \infty$, we have $\nu \ll \mathcal{H}^{n-4} \lfloor S$ (by Theorem A.21 (ii)). Hence, $\nu(U) = 0$. This means that $x \notin \mathrm{supp}(\nu) = \Gamma$, as we wanted. $\diamond$

Remark 4.13. When $n = 4$, Propositions 4.10 and 4.11 imply that $S = \Gamma$, so that the causes of the noncompactness along S only involves energy loss in this case. This is in accordance with the classical removable singularity theorem of Uhlenbeck (1982b). $\diamond$

4.2 Rectifiability of bubbling loci

As a first step towards understanding the noncompactness phenomenon involving energy loss, in this short section we show an important regularity result about the set Γ at which this phenomenon occurs.

Theorem 4.14 (Rectifiability of the bubbling locus). *The bubbling locus Γ is countably $\mathcal{H}^{n-4}$-rectifiable (cf. Definition A.40) and*

$$\nu = \Theta(\mu, \cdot)\mathcal{H}^{n-4} \lfloor \Gamma.$$

Proof. By Lemma 4.4 and Proposition 4.10, we know that

$$0 < \varepsilon_0 \leqslant \Theta^{n-4}(\nu, x) \leqslant e^{a\delta_0^2}\delta_0^{4-n}\Lambda < \infty, \quad \text{for } \mathcal{H}^{n-4} - \text{a.e. } x \in S.$$

Also, in general, $\Theta^*(\nu, \cdot) \leqslant \Theta(\mu, \cdot) < \infty$, which implies that $\nu \ll \mathcal{H}^{n-4} \lfloor S$ (cf. Theorem A.21 (ii)). Since $\Gamma = \mathrm{supp}(\nu)$, we get

$$0 < \varepsilon_0 \leqslant \Theta^{n-4}(\nu, x) \leqslant e^{a\delta_0^2}\delta_0^{4-n}\Lambda < \infty, \quad \text{for } \nu - \text{a.e. } x \in \Gamma.$$

Therefore, we can apply Theorem A.51 to conclude that Γ is countably $\mathcal{H}^{n-4}$-rectifiable and ν can be written as $\nu = \widetilde{\Theta}\mathcal{H}^{n-4} \lfloor \Gamma$, for some Borel measurable function $\widetilde{\Theta}$. In fact, since $\Theta^{n-4}(\mu, x) = \Theta^{n-4}(\nu, x)$ for $\mathcal{H}^{n-4}$-a.e. $x \in \Gamma$, we conclude that $\widetilde{\Theta}(x) = \Theta(\mu, x)$ for $\mathcal{H}^{n-4}$-a.e. $x \in \Gamma$. $\square$

Remark 4.15. Theorem 4.14 corresponds to Tian (2000, Proposition 3.3.3); Tian devotes the whole §3.3 of his paper for an independent proof of this result without relying on Preiss' theorem A.51. ◇

Remark 4.16. This theorem is trivial for $n = 4$: a set is countably $\mathcal{H}^0$-rectifiable if, and only if, it is at most countable, and according to Remark 3.50 this is indeed the case. In truth, as expected, the analysis of this chapter has content only when $n > 4$. ◇

Since $\mathcal{H}^{n-4}(S \setminus \Gamma) = 0$ (cf. Propositions 4.10 and 4.11), it follows that the blow-up locus itself is countably $\mathcal{H}^{n-4}$-rectifiable, and for $\mathcal{H}^{n-4}$-a.e. $x \in S$ the energy density lost by the sequence around the point x is measured by $\Theta^{n-4}(\mu, x)$.

Using Theorem A.50, we get the following consequence of Theorem 4.14:

Corollary 4.17. *At $\mathcal{H}^{n-4}$-a.e. $x \in \Gamma$, the bubbling locus has a well-defined tangent space $T_x \Gamma \subseteq T_x M$ and v has a unique tangent measure, i.e. the limit*

$$T_x v := \lim_{\lambda \to 0} \lambda^{4-n} (\exp_x \circ \tau_\lambda)^* v$$

exists and

$$T_x v = \Theta(\mu, x) \mathcal{H}^{n-4} \lfloor T_x \Gamma.$$

Here τ_λ denotes the scaling map on $T_x M$ taking v to λv.

Definition 4.18. We will say that $x \in \Gamma$ is a **smooth point** when the following holds:

(i) The tangent space $T_x \Gamma \subseteq T_x M$ is well-defined.

(ii) $x \notin \mathrm{sing}(\nabla)$.

By Theorem 4.14 and Proposition 4.10, it follows that the set of points in Γ which are not smooth in the above sense is $\mathcal{H}^{n-4}$-negligible.

4.3   Bubbling analysis

In this section we analyze the structure of ∇_i near smooth points of Γ when $i \gg 1$. We deduce the existence of non-trivial connections bubbling off transversely to Γ. The main reference for the approach here is Walpuski (2017c, §7), adapted to the context of Tian (2000, §3.2 - §4.1).

Fix a smooth point $x \in \Gamma$. Given a scale factor $\lambda > 0$, we define a rescaled sequence of connections on $T_x M$ by

$$\nabla_{i,x,\lambda} := (\exp_x \circ \tau_\lambda)^* \nabla_i, \tag{4.19}$$

with $\tau_\lambda(v) := \lambda v$ for all $v \in T_x M$. Then the $\nabla_{i,x,\lambda}$ are Yang–Mills connections with respect to rescaled metric

$$g_{x,\lambda} := \lambda^{-2} \exp_x^* g \tag{4.20}$$

on $T_x M$. We note further that $g_{x,\lambda}$ converges in C_{loc}^∞ to the flat metric $g_{x,0} = g|_{T_x M}$ on $T_x M$ as $\lambda \downarrow 0$. With the above notations, we introduce the following:

Definition 4.21 (Bubbling). A **bubble** (or a **bubbling connection**) B at $x \in \Gamma$ is a smooth Yang–Mills connection on the trivial G–bundle over $T_x M \cong \mathbb{R}^n$ which is invariant (modulo gauge) under translation with respect to the $(n-4)$–subspace $T_x \Gamma \subset T_x M$ and which is the limit of a *blowing-up* of the sequence $\{\nabla_i\}$ around the point x, i.e. there exists null-sequences[3] $\{u_i\} \subset T_x M$ and $\{r_i\} \subset \,]0, \tfrac{1}{2}[$ such that the blow-ups $\nabla_{i,x,r_i}(r_i^{-1} u_i + \cdot)$ converge in C_{loc}^∞ to B as $i \to \infty$.

We define the **energy** of the bubble B to be

$$\mathcal{E}(B) := \int_{T_x \Gamma^\perp} |F_B|^2.$$

Remark 4.22. Note that (modulo gauge) one may write a bubble B at x as the pullback to $T_x M$ of a Yang–Mills connection I on $T_x \Gamma^\perp \cong \mathbb{R}^4$, so that $\mathcal{E}(B) = \mathcal{YM}(I)$. In fact, it is common to not distinguish between B and I and refer to them indistinctly as a bubble. One says that I *bubbles off transversely* to Γ.

Our goal in this section is to show that part (iv) of Theorem B holds, i.e. we want to show that at any smooth point $x \in \Gamma$ there is a non-trivial bubble $B(x)$ whose energy is bounded by $\Theta(x)$.

We start with the following easy consequence of Corollary 4.17, which fixes a suitable null-sequence of scales $\{\lambda_i\}$ to work with.

Lemma 4.23 (Scale fixing). *If $x \in \Gamma$ is a smooth point, then we can find a null-sequence $\{\lambda_i\} \subseteq \,]0, 1[$ such that*

$$|F_{\nabla_{i,x,\lambda_i}}|^2 \mu_{g_{x,\lambda_i}} \rightharpoonup T_x \nu = \Theta(\mu, x) \mathcal{H}^{n-4} \lfloor T_x \Gamma. \tag{4.24}$$

---

[3] A sequence $\{x_i\}$ in a topological vector space X is called a *null-sequence* if it converges to $0 \in X$.

Proof. By Corollary 4.17, ν has a unique tangent measure $T_x \nu :=
\lim_{\lambda \downarrow 0} \lambda^{4-n}(\exp_x \circ \tau_\lambda)^* \nu = \Theta(\mu, x)\mathcal{H}^{n-4}\lfloor T_x \Gamma$. Since $x \notin \mathrm{sing}(\nabla)$, we have

$$\lim_{\lambda \downarrow 0} \lambda^{4-n}(\exp_x \circ \tau_\lambda)^* \nu = \lim_{\lambda \downarrow 0} \lambda^{4-n}(\exp_x \circ \tau_\lambda)^* \mu.$$

Thus, since $\mu_i \rightharpoonup \mu$,

$$T_x \nu = \lim_{\lambda \downarrow 0} \lim_{i \to \infty} \lambda^{4-n}(\exp_x \circ \tau_\lambda)^* \mu_i = \lim_{i \to \infty} \lambda_i^{4-n}(\exp_x \circ \tau_{\lambda_i})^* \mu_i,$$

for some null-sequence $\{\lambda_i\}$. This shows the claim since

$$\lambda_i^{4-n}(\exp_x \circ \tau_{\lambda_i})^* \mu_i = |F_{\nabla_{i,x,\lambda_i}}|^2 \mu_{g_{x,\lambda_i}}.$$

$$\square$$

Henceforth, we use the following notations. We write $N_x \Gamma := T_x \Gamma^\perp \subset
T_x M$ and use (z, w) to denote points in $T_x \Gamma \times N_x \Gamma = T_x M$. Moreover, we
shall work with generalized cubes of the form

$$Q_{r,s}(z_0, w_0) := B_r(z_0) \times B_s(w_0) \subset T_x \Gamma \times N_x \Gamma = T_x M.$$

The following is essentially Tian (2000, Lemma 4.1.2, p. 235).

Proposition 4.25 (Asymptotic translation invariance). *Let $x \in \Gamma$ be a smooth
point and let $\{\lambda_i\}$ be the null-sequence given by Lemma 4.23. Then after passing to a subsequence there is a null-sequence $\{z_i\} \subset B_1(0) \subset T_x \Gamma$ so that*

$$\lim_{i \to \infty} \sup_{s \leqslant 1} s^{4-n} \int_{Q_{s,1}(z_i,0)} \left| \frac{\partial}{\partial v} \lrcorner F_{\nabla_{i,x,\lambda_i}} \right|^2 dV_{g_{x,\lambda_i}} = 0. \qquad (4.26)$$

for any unit tangent vector $v \in T_x \Gamma$.

We shall split the proof of Proposition 4.25 into the following lemmas.

Lemma 4.27 (Tian (ibid., Lemma 3.3.2, p. 231)). *Under the hypothesis of
Proposition 4.25,*

$$\lim_{i \to \infty} \int_{Q_{2,1}(0,0)} \left| \frac{\partial}{\partial v} \lrcorner F_{\nabla_{i,x,\lambda_i}} \right|^2 dV_{g_{x,\lambda_i}} = 0.$$

Proof. Fix $g|_{T_x M}$–orthogonal coordinates $\{y_l\}_{l=1}^{n-4}$ on $T_x \Gamma$, with ∂_{y_l} having length 4. Let ∂_{ρ_i} denote the radial vector field emanating from the point ∂_{y_l} associated with the metric g_{x,λ_i}. Then the monotonicity formula (Theorem 3.24) implies that for $0 < s \leq r$

$$\int_{B_r(\partial_{y_l}) \setminus B_s(\partial_{y_l})} e^{a\lambda_i^2 \rho_i^2} \rho_i^{4-n} \left| \partial_{\rho_i} \lrcorner F_{\nabla_{i,x,\lambda_i}} \right|^2$$
$$\leq e^{a\lambda_i^2 r^2} r^{4-n} \int_{B_r(\partial_{y_l})} \left| F_{\nabla_{i,x,\lambda_i}} \right|^2 - e^{a\lambda_i^2 s^2} s^{4-n} \int_{B_s(\partial_{y_l})} \left| F_{\nabla_{i,x,\lambda_i}} \right|^2.$$

Now note that as $i \to \infty$ the two terms of the right-hand side both converge to $\Theta(\mu, x)$ by Lemma 4.23. Since $Q_{2,1}(0,0) \subset B_8(\partial_{y_l}) \setminus B_1(\partial_{y_l})$, we get

$$\lim_{i \to \infty} \int_{Q_{2,1}(0,0)} \left| \partial_{\rho_i} \lrcorner F_{\nabla_{i,x,\lambda_i}} \right|^2 dV_{g_{x,\lambda_i}} = 0.$$

Furthermore, at the origin the ∂_{ρ_i} generate $T_x \Gamma$ and, as the metrics g_{x,λ_i} converge to $g|_{T_x M}$, we may state the result in terms of it. The lemma is proved. $\qquad\square$

Lemma 4.28. *Under the hypothesis of Proposition 4.25, for $\mathcal{H}^{n-4}$–a.e. $z \in B_1(0) \subset T_x \Gamma$ one has*

$$\lim_{i \to \infty} \sup_{s \leq 1} s^{4-n} \int_{Q_{s,1}(z,0)} \left| \frac{\partial}{\partial v} \lrcorner F_{\nabla_{i,x,\lambda_i}} \right|^2 dV_{g_{x,\lambda_i}} = 0.$$

Proof. Define $f_i : B_2(0) \subset T_x \Gamma \to [0, \infty)$ by

$$f_i(z) := \int_{B_1(0) \subset N_x \Gamma} \left| \partial_v \lrcorner F_{\nabla_{i,x,\lambda_i}} \right|^2 (z, \cdot)$$

and denote by $M f_i : B_1(0) \subset T_x \Gamma \to [0, \infty)$ the Hardy–Littlewood maximal function associated with f_i:

$$M f_i(z) := \sup_{s \leq 1} s^{4-n} \int_{B_s(z) \subset T_x \Gamma} f_i.$$

We then want to show that the set

$$A := \{ z \in B_1(0) : \liminf_{i \to \infty} M f_i(z) > 0 \}$$

is such that $\mathcal{H}^{n-4}(A) = 0$. For each $j \in \mathbb{N}$, define

$$A_{i,j} := \{ z \in B_1(0) : M f_i(z) \geq j^{-1} \}.$$

Then we can write

$$A = \bigcup_{j \geqslant 1} \bigcup_{n \geqslant 1} \bigcap_{i \geqslant n} A_{i,j}.$$

For each j, on the one hand, by the weak-type L^1 estimate for the maximal operator, we have

$$\mathcal{H}^{n-4}(A_{i,j}) \lesssim j \, \|f_i\|_{L^1}.$$

On the other hand, by Lemma 4.27, it follows that $\|f_i\|_{L^1} \to 0$ as $i \to \infty$. Therefore, for each j and n we get

$$\mathcal{H}^{n-4}\left(\bigcap_{i \geqslant n} A_{i,j}\right) = 0,$$

which in turn implies that $\mathcal{H}^{n-4}(A) = 0$ by monotone convergence. This completes the proof of the lemma. $\qquad\square$

Proof of Proposition 4.25. By Lemma 4.28, for each $j \in \mathbb{N}$ we can find $z_j \in B_{1/j}(0) \subset T_x\Gamma$ such that

$$\lim_{i \to \infty} \sup_{s \leqslant 1} s^{4-n} \int_{Q_{s,1}(z_j,0)} \left|\frac{\partial}{\partial v} \lrcorner \, F_{\nabla_{i,x,\lambda_i}}\right|^2 \, dV_{g_{x,\lambda_i}} = 0.$$

The conclusion then follows by applying a standard diagonal sequence argument. $\qquad\square$

The next proposition is the last preparation for the main theorem of this section.

Proposition 4.29 (Bubble detection). *Let $x \in \Gamma$ be a smooth point, $\{\lambda_i\}$ be the null-sequence given by Lemma 4.23 and $\{z_i\} \subset T_x\Gamma$ the null-sequence given by Proposition 4.25. Then there exists a null-sequence $\{\delta_i\} \subset \,]0, \frac{1}{2}[$ such that, for each $i \gg 1$,*

$$\max_{w \in \overline{B}_{\frac{1}{2}}(0)} \delta_i^{4-n} \int_{B_{\delta_i}(z_i,w)} \left|F_{\nabla_{i,x,\lambda_i}}\right|^2 \, dV_{g_{x,\lambda_i}} = \frac{\varepsilon_0}{8}. \tag{4.30}$$

Moreover, if $w_i \in \overline{B}_{\frac{1}{2}}(0)$ denotes a point at which this maximum is achieved, then $\{w_i\}$ is a null-sequence.

Proof. On the one hand, by Lemma 4.23,

$$\liminf_{i\to\infty} \max_{w\in\overline{B}_{\frac{1}{2}}(0)} \delta^{4-n} \int_{B_\delta(z_i,w)} \left|F_{\nabla_{i,x,\lambda_i}}\right|^2 dV_{g_{x,\lambda_i}} = \Theta(x) \geq \varepsilon_0,$$

for all $\delta > 0$. On the other hand, for fixed $i \in \mathbb{N}$ and $w \in \overline{B}_{\frac{1}{2}}(0)$ by smoothness one has

$$\lim_{\delta\downarrow 0} \delta^{4-n} \int_{B_\delta(z_i,w)} \left|F_{\nabla_{i,x,\lambda_i}}\right|^2 dV_{g_{x,\lambda_i}} = 0.$$

Therefore, we can find a null-sequence $\{\delta_i\} \subset \,]0, \frac{1}{2}[$ such that (4.30) holds.

As for the last assertion, note that if we could find $\sigma > 0$ so that $\{w_i\} \subset \overline{B}_{\frac{1}{2}}(0) \setminus B_\sigma(0)$, then the density of $T_x\nu$ at $(0, w_i)$ would be positive, contradicting Lemma 4.23. $\qquad\square$

Now we can state and prove the main theorem of this section (cf. Tian (2000, Proposition 4.1.1, p. 235)).

Theorem 4.31 (Bubbling). *Let $x \in \Gamma$ be a smooth point, i.e. $T_x\Gamma$ exists and $x \notin \mathrm{sing}(\nabla)$. Then there exists a non-trivial bubble $B(x)$ at x whose energy is bounded by $\Theta(x)$. More precisely, there is a non-flat Yang–Mills connection $I(x)$ on $N_x\Gamma$ satisfying*

$$\mathcal{YM}(I(x)) \leq \Theta(\mu, x) \tag{4.32}$$

and whose pullback $B(x)$ to T_xM is (modulo gauge) the limit of a blowing-up of $\{\nabla_i\}$ around x (cf. Definition 4.21).

Proof. Let $\{\lambda_i\}, \{\delta_i\} \subset \,]0, \frac{1}{2}[$ and $\{(z_i, w_i)\} \subset T_x\Gamma \times N_x\Gamma = T_xM$ be the null-sequences given by the previous results of this section. Define the sequence of blow-ups

$$\widetilde{\nabla}_i(\cdot) := \nabla_{i,x,\delta_i\lambda_i}(\delta_i^{-1}(z_i, w_i) + \cdot) = \nabla_{i,x,r_i}(r_i^{-1}u_i + \cdot),$$

where $u_i := \lambda_i(z_i, w_i)$ and $r_i := \delta_i\lambda_i$. By construction

$$\max_{w\in\overline{B}_{(\frac{1}{2}-|w_i|)\delta_i}(0)} \int_{B_1(0,w)} \left|F_{\widetilde{\nabla}_i}\right|^2 dV_{g_{x,0}} = \frac{\varepsilon_0}{8}, \tag{4.33}$$

with the maximum achieved at $w = 0$. It then follows from Theorem 3.33, and standard elliptic techniques, that $\widetilde{\nabla}$ converges in C^∞_{loc} to a Yang–Mills connection $B(x)$ over $B_1(0) \times N_x\Gamma$ with the flat metric $g_{x,0}$. Moreover, by Proposition 4.25, one has

$$v\lrcorner F_{B(x)} = 0, \quad \forall v \in T_x\Gamma.$$

This implies that $B(x)$ is gauge-equivalent to the pullback of a Yang–Mills connection $I(x)$ on $N_x\Gamma$. The convergence together with (4.33) shows that $I(x)$ is non-flat and that (4.32) holds. $\square$

Remark 4.34. Taking into account the results that have been proven so far, this completes the proof of Theorem A stated in the introduction of this work.

Given the above bubble-detection result at a smooth point $x \in \Gamma$, one is then tempted to ask about whether $\Theta(x)$ can actually be written as a finite sum of energies of non-flat Yang–Mills connections on $\mathbb{R}^4 \cong N_x\Gamma$ (or, by transforming conformally, Yang–Mills connections on S^4) arising as bubbles at x. Indeed, we know that each concentration generating a bubble like in Theorem 4.31 has a cost of energy bounded from below by the Uhlenbeck-constant of $\mathbb{R}^4$ (the ε_0 of Theorem 3.33 applied to the flat $\mathbb{R}^4$), thus the inequality (4.32) shows that the number of possible distinct bubbles is uniformly bounded.

This *energy identity* question was first positively answered by Rivière Rivière (2002) assuming an uniform L^1 hessian bound on the curvatures $\|\nabla^2 F_\nabla\|_{L^1(M)} \leqslant C(\Lambda, G, M)$. More recently, Naber–Valtorta A. Naber and Valtorta (2016) proved the energy identity in tandem with the fact that such hessian bound always holds automatically. This is a deep result and we will just cite it below as follows:

Theorem 4.35 (A. Naber and Valtorta (ibid.)). *For $\mathcal{H}^{n-4}-a.e.$ $x \in \Gamma$, there exists a finite collection of distinct bubbles $B_1(x),\dots, B_l(x)$ at x such that*

$$\Theta(x) = \sum_{j=1}^{l} \mathcal{E}(B_j(x)).$$

This means that there is no loss of energy between bubbles; this is called the *no neck property*. This property was previously established in four dimensions for sequences of ASD connections an was an important step towards the compactification of the moduli space of instantons on compact 4−manifolds, cf. Donaldson and Kronheimer (1990).

4.4   Blow-up loci of instantons and calibrated geometry

Throughout this section we assume further that M is endowed with a smooth $(n-4)-$form Ξ which is a calibration on (M, g) and that $\{\nabla_i\}$ is a sequence of $\Xi-$ASD instantons on the $G-$bundle E (cf. Definition 2.80). Our aim is to show stronger conclusions in this setting for the results derived so far for general Yang–Mills connections. We remark that in case M is closed (compact without boundary) and $G = \mathrm{SU}(r)$, it follows from Proposition 2.84 that we have the a priori energy bound $\mathcal{YM}(\nabla_i) = 8\pi^2 \langle c_2(E) \cup [\Xi], [M] \rangle$.

We begin with the following simple

Lemma 4.36. *In the above setting, the limit connection ∇ on $E|_{M \setminus S}$ is in fact a Ξ-ASD instanton on $M \setminus S$.*

Proof. By assumption, there exists a sequence of gauge-transformations $\{g_i\} \subseteq \mathcal{G}(E|_{M \setminus S})$, $S = S(\{\nabla_i\})$, such that $g_i^* \nabla_i$ converges to ∇ in C^∞_{loc}-topology outside S. In particular, it follows that $\mathrm{tr}(F_{\nabla_i})$ converges to $\mathrm{tr}(F_\nabla)$ in C^∞_{loc}-topology outside S, and $\mathrm{tr}(F_\nabla)$ is harmonic o $M \setminus S$. Since $*_\Xi = *(\Xi \wedge \cdot)$ is clearly a continuous operator with respect to the C^∞_{loc}-topology, and each ∇_i is Ξ-ASD, the result follows from the equivariance of $*_\Xi$. $\square$

We actually know more about the bubbling connections prescribed in Proposition 4.31, cf. Tian (2000, Theorem 4.2.1).

Proposition 4.37 (Bubbling Ξ-ASD connections). *Let $B(x)$ be a bubble connection at a smooth point $x \in \Gamma$ (cf. Proposition 4.31). Then $B(x)$ is a non-flat Ξ_x-ASD instanton on $(T_x M, g_{x,0})$ with $\mathrm{tr}(F_{B(x)}) = 0$.*

Proof. Recall that $B_i(x) := \sigma_i^* \exp_x^* \nabla_i$, where $\sigma_i : T_x M \to T_x M$ is of the form $v \mapsto \lambda_i(z_i, w_i) + \lambda_i \delta_i v$, with $(z_i, w_i) \to 0$ and $\lambda_i, \delta_i \to 0$ as $i \to \infty$. Since $\mathrm{tr}(F_{\nabla_i}) \to \mathrm{tr}(F_\nabla)$ uniformly on compact subsets outside S as $i \to \infty$, it follows that $\mathrm{tr}(F_{B_i(x)}) \to 0$ uniformly on compact subsets as $i \to \infty$.

On the other hand, note that $B_i(x)$ is $(\lambda_i \delta_i)^{4-n} \sigma_i^* \exp_x^* \Xi$-ASD with respect to the metric $g_{x, \lambda_i \delta_i}$. Moreover, since σ_i converges to zero, we have that $\sigma_i^* \exp_x^* \Xi \to \Xi_x$ as $i \to \infty$.

In conclusion, recalling that $g_{x, \lambda_i \delta_i} \to g_{x,0}$ as $i \to \infty$, it follows that the limit connection $B(x)$ is Ξ_x-ASD with respect to the flat metric $g_{x,0}$, and $\mathrm{tr}(F_{B(x)}) = 0$. $\square$

The combination of this last result with Proposition 2.88 immediately yields:

Corollary 4.38. *At each smooth point $x \in \Gamma$ there is a choice of orientation on $(T_x \Gamma, g|_{T_x \Gamma})$ with respect to which it is calibrated by Ξ_x. Furthermore, if $B(x)$ is a bubbling connection at x then $B(x)$ is gauge-equivalent to the pullback to $T_x M$ of a non-trivial ASD instanton $I(x)$ on $(N_x \Gamma, g|_{N_x \Gamma})$ with respect to the induced orientation $*\Xi_x|_{N_x \Gamma}$.*

Now, combining Corollary 4.38 with the energy quantization of ASD instantons on S^4 (cf. §1.5), the following is immediate from Theorem 4.35:

Theorem 4.39. *Suppose* $G = \mathrm{SU}(r)$. *Then* $\Theta(x) \in 8\pi^2 \mathbb{Z}$ *for all smooth points* $x \in \Gamma$.

Finally, we conclude the proof of Theorem B (stated in the introduction), by proving the following (cf. Tian (2000, Theorem 4.3.2, eq. (4.2.5))):

Theorem 4.40. *Suppose* $G = \mathrm{U}(r)$ *and let* $C(\Gamma, \Theta) \in \mathscr{D}_{n-4}(M)$ *be defined by*

$$C(\Gamma, \Theta)(\varphi) := \frac{1}{8\pi^2} \int_\Gamma \langle \varphi, \Xi |_\Gamma \rangle \Theta d\left(\mathcal{H}^{n-4} \lfloor \Gamma\right), \quad \forall \varphi \in \mathscr{D}^{n-4}(M).$$

Then $C(\Gamma, \Theta)$ *satisfy the following weak convergence of currents:*

$$c_2(\nabla_i) \rightharpoonup c_2(\nabla) + C(\Gamma, \Theta). \tag{4.41}$$

Remark 4.42. Here we let $c_2(\nabla_{(i)}) \in \mathscr{D}_{n-4}(M)$ be defined by

$$c_2(\nabla_{(i)})(\varphi) := \frac{1}{8\pi^2} \int_M \varphi \wedge \left(\mathrm{tr}(F_{\nabla_{(i)}} \wedge F_{\nabla_{(i)}}) - \mathrm{tr}(F_{\nabla_{(i)}}) \wedge \mathrm{tr}(F_{\nabla_{(i)}})\right),$$

for all $\varphi \in \mathscr{D}^{n-4}(M)$. Now note that

$$\mathrm{tr}(F_{\nabla_{(i)}} \wedge F_{\nabla_{(i)}}) = \mathrm{tr}(F^0_{\nabla_{(i)}} \wedge F^0_{\nabla_{(i)}}) + \frac{1}{r}\mathrm{tr}(F_{\nabla_{(i)}}) \wedge \mathrm{tr}(F_{\nabla_{(i)}}).$$

Since $\mathrm{tr}(F_{\nabla_i})$ converges to $\mathrm{tr}(F_\nabla)$ in C^∞_{loc}−topology outside S, we get:

$$8\pi^2 \left[c_2(\nabla_i) - c_2(\nabla)\right](\varphi) = \lim_{i \to \infty} \int_M \varphi \wedge \left(\mathrm{tr}(F^0_{\nabla_i} \wedge F^0_{\nabla_i}) - \mathrm{tr}(F^0_\nabla \wedge F^0_\nabla)\right), \tag{4.43}$$

for all $\varphi \in \mathscr{D}^{n-4}(M)$. In particular, equation (4.41) is equivalent to

$$\frac{1}{8\pi^2} \lim_{i \to \infty} \int_M \varphi \wedge \mathrm{tr}(F^0_{\nabla_i} \wedge F^0_{\nabla_i}) = \frac{1}{8\pi^2} \int_M \varphi \wedge \mathrm{tr}(F^0_\nabla \wedge F^0_\nabla) + C(\Gamma, \Theta)(\varphi), \tag{4.44}$$

for all $\varphi \in \mathscr{D}^{n-4}(M)$. Of course, if $G = \mathrm{SU}(r)$ then $F^0_{\nabla_{(i)}} = F_{\nabla_{(i)}}$ and we get precisely (B1). Finally, in case M is closed note that we can apply equation (4.44) in $8\pi^2 \Xi$ to deduce the L^2−**energy conservation:**

$$\lim_{i \to \infty} \int_M |F_{\nabla_i}|^2 dV_g = \int_M |F_\nabla|^2 dV_g + \int_\Gamma \Theta d\left(\mathcal{H}^{n-4}\lfloor \Gamma\right),$$

i.e. $\int_\Gamma \Theta d\mathcal{H}^{n-4}\lfloor \Gamma$ is precisely the L^2−energy lost by the sequence $\{\nabla_i\}$ along Γ as $i \to \infty$. $\Diamond$

Proof of Theorem 4.40. By Remark 4.42, it suffices to show that (4.44) holds. In fact, the $C_{\mathrm{loc}}^{\infty}$–convergence $\mathrm{tr}(F_{\nabla_i}) \to \mathrm{tr}(F_{\nabla})$ outside S allow us to suppose, without loss of generality, that $G = \mathrm{SU}(r)$.

For each $i \in \mathbb{N}$, define the current $T_i \in \mathscr{D}_{n-4}(M)$ given by

$$T_i := c_2(\nabla_i) - c_2(\nabla).$$

Since $\mathcal{YM}(\nabla_i) \leq \Lambda$, note that the total mass $\mathbf{M}(T_i)$ of T_i is uniformly bounded: indeed, for any $\varphi \in \mathscr{D}^{n-4}(M)$ with $\|\varphi\|_{C^0} \leq 1$ we have:

$$|T_i(\varphi)| \lesssim \|F_{\nabla_i}\|_{L^2}^2 + \|F_{\nabla}\|_{L^2}^2 \lesssim \Lambda.$$

Therefore, by Lemma A.63, up to taking a subsequence if necessary we may suppose that $T_i \rightharpoonup T$ for some $T \in \mathbf{M}_{n-4,\mathrm{loc}}(M)$. We then need to prove that $T = C(\Gamma, \Theta)$, and in order to do so we shall first verify that T satisfies the hypothesis of Theorem A.68.

By the weak convergence, it follows that $\mathbf{M}(T) \leq \liminf \mathbf{M}(T_i) \lesssim \Lambda$. Moreover, note that $\|\partial T\|(W) < \infty$ for all $W \Subset U$; indeed, $\partial T = -\partial c_2(\nabla)$ and

$$\partial T(\varphi) = -\int_M d\varphi \wedge \mathrm{tr}(F_{\nabla} \wedge F_{\nabla}) \leq \left(\sup_W |d\varphi| \right) \Lambda < \infty,$$

for all $\varphi \in \mathscr{D}^{n-5}(M)$ with $\mathrm{supp}(\varphi) \subseteq W$. Hence, $T \in \mathbf{N}_{n-4,\mathrm{loc}}(M)$.

Now we show that

$$\Theta^{*n-4}(\|T\|, x) > 0 \quad \text{for } \|T\| - \text{a.e. } x \in M. \tag{4.45}$$

First note that the convergence (modulo gauge) of ∇_i to ∇ in $C_{\mathrm{loc}}^{\infty}$ away from S immediately implies that $\mathrm{supp}(T) \subseteq S$. In particular, we have that $\mathrm{supp}(\|T\|) \subseteq S$. We claim further that

$$\|T\| \ll \mathcal{H}^{n-4}\lfloor \Gamma. \tag{4.46}$$

Indeed, let $0 < r \leq \delta_0$ and $x \in \Gamma$. Then, whenever $\varphi \in \mathscr{D}^{n-4}(M)$ is such that $\|\varphi\|_{C^0} \leq 1$ and $\mathrm{supp}(\varphi) \subseteq B_r(x)$ we have

$$|T_i(\varphi)| \lesssim \left(\int_{B_r(x)} |F_{\nabla_i}|^2 dV_g + \int_{B_r(x)} |F_{\nabla}|^2 dV_g \right).$$

Recalling that $\Theta(\mu, \cdot)$ is bounded and $\mu_i \rightharpoonup \mu = |F_{\nabla}|^2 \mu_g + \Theta(\mu, \cdot)\mathcal{H}^{n-4}\lfloor \Gamma$, it follows that

$$\|T\|(B_r(x)) \lesssim \left(\int_{B_r(x)} |F_{\nabla}|^2 dV_g + \mathcal{H}^{n-4}(\Gamma \cap B_r(x)) \right).$$

Now Proposition 4.10 ensures that $\Theta^*(\nabla, x) = 0$ for $\mathcal{H}^{n-4}$−a.e. $x \in M$, and since $\mathcal{H}^{n-4}(\Gamma) < \infty$ it follows from Theorem A.23 that $\Theta^{*n-4}(\Gamma, x) \leqslant 1$ for $\mathcal{H}^{n-4}$−a.e. $x \in \Gamma$. Thus

$$\Theta^{*n-4}(\|T\|, x) \lesssim 1 \quad \text{for } \mathcal{H}^{n-4}\text{−a.e. } x \in \Gamma.$$

Therefore, the claim (4.46) follows by Theorem A.21.

Now let $f \in \mathscr{D}(M)$ and take $\varphi = f\,\Xi$; then:

$$\begin{aligned}
T(f\,\Xi) &= \lim_{i \to \infty} T_i(f\,\Xi) \\
&= \frac{1}{8\pi^2} \lim_{i \to \infty} \int_M f \left(\mathrm{tr}(F_{\nabla_i} \wedge F_{\nabla_i}) - \mathrm{tr}(F_\nabla \wedge F_\nabla) \right) \wedge \Xi \\
&= \frac{1}{8\pi^2} \lim_{i \to \infty} \int_M f \left(-\mathrm{tr}(F_{\nabla_i} \wedge *F_{\nabla_i}) + \mathrm{tr}(F_\nabla \wedge *F_\nabla) \right) \\
&= \frac{1}{8\pi^2} \lim_{i \to \infty} \int_M f \left(|F_{\nabla_i}|^2 - |F_\nabla|^2 \right) \\
&= \frac{1}{8\pi^2} \int_\Gamma f\,\Theta(\mu, \cdot)\mathrm{d}\left(\mathcal{H}^{n-4}\lfloor\Gamma \right).
\end{aligned} \tag{4.47}$$

Thus, if $x \in \Gamma$ and $r > 0$, we get

$$\|T\|(B_r(x)) \geqslant \frac{1}{8\pi^2} \int_{B_{\frac{r}{2}}(x) \cap \Gamma} \Theta(\mu, \cdot)\mathrm{d}\left(\mathcal{H}^{n-4}\lfloor\Gamma \right) \geqslant \frac{\varepsilon_0}{8\pi^2} \mathcal{H}^{n-4}(\Gamma \cap B_{\frac{r}{2}}(x)).$$

Since $\mathcal{H}^{n-4}(\Gamma) < \infty$, by Theorem A.23 we know that $\Theta^{*n-4}(\Gamma, x) \geqslant 2^{4-n} > 0$ for $\mathcal{H}^{n-4}$−a.e. $x \in \Gamma$. Thus, it follows that $\Theta^{*n-4}(\|T\|, x) > 0$ for $\mathcal{H}^{n-4}$−a.e. $x \in \Gamma$. Together with (4.46) this implies (4.45).

Now we may apply Theorem A.68 to conclude that we can find a triple (Γ', Θ', ξ) such that

$$T(\varphi) = \frac{1}{8\pi^2} \int_{\Gamma'} \langle\varphi, \xi\rangle \Theta'\mathrm{d}\left(\mathcal{H}^{n-4}\lfloor\Gamma' \right), \quad \forall\varphi \in \mathscr{D}^{n-4}(M)$$

where

1. $\Gamma' \subseteq M$ is $\mathcal{H}^{n-4}$−measurable and countably $\mathcal{H}^{n-4}$−rectifiable;

2. $\Theta' \colon \Gamma \to [0, \infty[$ is locally $\mathcal{H}^{n-4}$−integrable;

3. $\xi \colon \Gamma \to \Lambda^k TM$ is $\mathcal{H}^{n-4}$−measurable and such that $\xi(x)$ orients the approximate $(n - 4)$−tangent space $T_x\Gamma'$ for $\mathcal{H}^{n-4}$−a.e. $x \in \Gamma'$.

In particular, for every $f \in \mathscr{D}(M)$ we have

$$T(f\,\Xi) = \frac{1}{8\pi^2} \int_{\Gamma'} f\,\langle\Xi,\xi\rangle\Theta'\,\mathrm{d}\left(\mathcal{H}^{n-4}\lfloor\Gamma'\right).$$

Comparing with (4.47), we conclude that $\Gamma = \Gamma'$ and $\langle\Xi,\xi\rangle\Theta' = \Theta(\mu,\cdot)$. Finally, since $\Xi|_\Gamma$ is one of the volume forms of Γ we get $\Theta' = \Theta$, so that $T = C(\Gamma,\Theta)$. $\qquad\square$

Remark 4.48. Note that in the situation of Theorem 4.40 it follows directly that

$$\partial C = 0 \quad \Longleftrightarrow \quad \partial c_2(\nabla) = 0, \tag{4.49}$$

in which case C would be a Ξ–calibrated *cycle*, therefore the bubbling locus would be mass-minimizing by Proposition 2.59, and its components Ξ–submanifolds except for singular sets of Hausdorff codimension at least 2 by Theorem 2.60. Nevertheless, we note that this may not be true in general, and refer the reader to Petrache and Rivière (2017, §1.12.4).

Now that we finished the proof of both Theorems A and B, we note that we get obvious important corollaries in each special case where (M,g) is a Riemannian manifold with one of the special holonomy groups $\mathrm{Hol}(g) = \mathrm{U}(m)$, G_2 or $\mathrm{Spin}(7)(\supseteq \mathrm{SU}(4))$. In particular, in the Kähler context, we have the following:

Theorem 4.50. *Let (Z,ω) be a compact Kähler m–fold, and let $\{\nabla_i\}$ be a sequence of Hermitian–Yang–Mills connections with uniformly bounded L^2–energy on a G–bundle E over Z. Then, by taking a subsequence if necessary, ∇_i converges modulo gauge to a Hermitian–Yang–Mills connection ∇ outside the bubbling locus (Γ,Θ), where $\Gamma = \cup_\alpha \Gamma_\alpha$ and $\Theta|_{\Gamma_\alpha} = 8\pi^2 m_\alpha$, where each Γ_α is a codimension-4 complex subvariety in Z and m_α is a positive integer. Moreover,*

$$c_2(\nabla_i)(\varphi) \rightharpoonup c_2(\nabla)(\varphi) + \sum_\alpha m_\alpha \int_{\Gamma_\alpha} \varphi.$$

Proof. By Lemma 2.97 we can apply Theorems A and B to the sequence $\{\nabla_i\}$. Then, using Corollary 2.62 and a result of Harvey–Shiffman Harvey and Shiffman (1974, Theorem 2.1) the theorem follows. $\qquad\square$

Interesting current problems stem from trying to see Tian's result in practice, to understand the regularity of the limiting configurations (∇,Γ,Θ) and also to reverse the process: given a 'generic' Ξ–calibrated submanifold $\Gamma \hookrightarrow M$ of a closed special holonomy Riemannian manifold (M^n,g,Ξ), and a 'generic' connection ∇ on a $G-$bundle over M, one may ask when does Γ appear as the bubbling locus of a sequence of $\Xi-$ASD instantons, smoothly converging to ∇ outside Γ.

Concerning explicit non-trivial examples of instanton bubbling, and removable singularity phenomena in the limit, the reader may consult the recent works of Lotay–Oliveira Lotay and Oliveira (2018) for the G_2 case and Clarke–Oliveira Clarke and Oliveira (2019) for the Spin(7) case.

As to the regularity of limiting connections, and an analytical framework analogous to the space of Sobolev connections, yet adapted to this higher dimensional context, the recent work of Petrache–Rivière Petrache and Rivière (2018) introduces a space of so-called 'weak connections'. They prove a weak sequential closure property, under Yang–Mills energy control, and they establish a strong approximation property of any weak connection by smooth connections away from polyhedral sets of codimension 5. In direct relation to this theory, partial regularity results and removable singularity theorems for general *stationary Yang–Mills* [4] connections, satisfying some approximability property, see Meyer and Rivière (2003) and Tao and Tian (2004).

Now, in the direction of the 'reverse process' of bubbling, Walpuski (2017a,b) gave sufficient conditions for an unobstructed associative (resp., Cayley) submanifold in a G_2-(resp., Spin(7)–)manifold to appear as the bubbling locus of a sequence of G_2-(resp. Spin(7)–)instantons, related to the existence of a Fueter section of a bundle of ASD instanton moduli spaces over said submanifold. Thus associative (resp. Cayley) submanifolds, and connections on them, arise as building blocks for constructing G_2-(resp., Spin(7)–)instantons by gluing methods. This has been successfully implemented on both Joyce's construction and Kovalev-CNHP twisted connected sums Sá Earp and Walpuski (2015) and Walpuski (2013a).

One can also attempt to construct invariants of G_2-manifolds by "counting" G_2-instantons and associative submanifolds together, as first suggested in Donaldson and Thomas (1998). In fact, Donaldson and Segal (2011) proposed a conjectural programme to define Casson-type invariants of G_2-instantons, which would hopefully be unchanged under deformations of the G_2-structure, and would be analogues of the invariants of Calabi–Yau 3–folds defined by Donaldson and Thomas (1998), see also Thomas (1997). They observed that the naive count of G_2-instantons on a compact G_2-manifold might not produce a deformation-invariant number, but rather this number will jump in a finite number of points as one changes the G_2-metric in a 1–parameter family. These jumps are closely related to degenerations of G_2-instantons to Fueter

---

[4]A connection $\nabla \in \mathcal{A}^{1,2} \cap \mathcal{A}^{0,4}$ is said to be *stationary Yang–Mills* if it is weak Yang–Mills and satisfies equation (3.15).

sections supported on certain associative submanifolds. Thus, completing such programme and defining invariants of a G_2−manifold (Y^7, ϕ) that remain unchanged under deformations of ϕ, would require the inclusion of 'compensation terms', counting solutions of some 'Dirac type' equation on associative 3−folds in Y, to balance out the bubbling of G_2−instantons. At the time of writing, this is mostly conjectural and currently under investigation, see e.g. Doan (2019), Doan and Walpuski (2017), Haydys (2012, 2017), Haydys and Walpuski (2015), Joyce (2017), and Walpuski (2013b).

Geometric
measure
theory

A

We collect here some basic facts from Geometric Measure Theory (GMT) which are evoked in Chapter 4. We stress that we have no intention to make a complete systematic exposition here, but rather just organise the main definitions and results, and fix some notation and conventions. We will therefore omit almost all proofs and refer the reader to standard texts. Good references for the material in this appendix are notes by Simon (1983), the classic Federer (1969), and the more recent books Mattila and Falconer (1996) and De Lellis (2008).

Notation. Throughout X will denote a metric space with distance function d. For any subset $A \subseteq X$, we denote by $\overline{A}$, $\mathring{A}$ and ∂A, respectively, the topological closure, interior and boundary of A. For each $x \in X$ and $r \in \mathbb{R}_+$, we write $B_r(x)$, $\overline{B}_r(x)$ and $\partial B_r(x)$ to denote, respectively, the *open* ball with center x and radius r, its closure and its boundary. If B is an open (resp. closed) ball in X of center x and radius r, then for each positive real number $\lambda > 0$ we write λB for the open (resp. closed) ball in X of center x and radius λr. The distance between two subsets $A, B \subseteq X$ is denoted by $d(A, B)$, and the diameter of A is denoted by $\mathrm{diam}(A)$. Finally, we shall use the extended real number system $\overline{\mathbb{R}} = \mathbb{R} \cup \{\infty, -\infty\}$ with the obvious ordering and aritheoremetical operations partially extended, e.g., as in Folland (2013, p. 11).

A.1   Basic concepts

Definition A.1 (Measures, $\sigma-$additivity and measurable sets). A **measure** (or *outer* measure) on X is a set function $\mu\colon 2^X \to [0, \infty]$ satisfying the following conditions:

  (i) $\mu(\emptyset) = 0$;

  (ii) (Monotonicity) $A \subseteq B,\ A, B \subseteq X \Rightarrow \mu(A) \leqslant \mu(B)$;

  (iii) (Subadditivity) For each countable collection $\{A_i\}_{i \in \mathbb{N}} \subseteq X$,

$$\mu\left(\bigcup_{i=1}^{\infty} A_i\right) \leqslant \sum_{i=1}^{\infty} \mu(A_i).$$

Given a family $\mathcal{F}$ of subsets of X, we say that μ is $\sigma-$**additive on** $\mathcal{F}$ whenever

$$\mu\left(\bigcup_{i=1}^{\infty} A_i\right) = \sum_{i=1}^{\infty} \mu(A_i),$$

for each countable collection of *disjoint* sets $\{A_i\}_{i \in \mathbb{N}} \subseteq \mathcal{F}$.

Finally, we say that a subset $A \subseteq X$ is $\mu-$**measurable** if

$$\mu(E) = \mu(E \setminus A) + \mu(E \cap A), \quad \text{for each } E \subseteq X.$$

We denote by $\mathcal{M}_\mu$ the collection of all $\mu-$measurable sets.

Definition A.2. Let μ be a measure on X. We define the **support** of μ to be the following closed subset of X:

$$\operatorname{supp}(\mu) := X \setminus \bigcup \{U \subseteq X : U \text{ is open and } \mu(U) = 0\}.$$

Given a measure μ on X, a sentence of the form

$$\text{``(...) holds for } \mu-\text{almost every point } x \in X\text{''}$$

or, briefly,

$$\text{``(...) holds for } \mu-\textbf{a.e. } x \in X\text{''}$$

means that the subset, say A, of X for which (...) doesn't hold is a $\mu-$**negligible set**, i.e. $\mu(A) = 0$.

Recall that a $\sigma-$**algebra** Σ on a set Y is a collection of subsets of Y, containing the empty set $\emptyset$ and Y itself, that is closed under the set operations of taking complements and countable unions. When Y is a topological space, the smallest $\sigma-$algebra $\mathcal{B}(Y)$ containing the topology of Y is called the **Borel** $\sigma-$algebra and its elements are the **Borel sets**.

In the next result we collect some well-known basic facts about general measures (see e.g. Folland (2013, §1)).

Theorem A.3. *If μ is a measure on X, then $\mathcal{M}_\mu$ is a $\sigma-$algebra on X. Moreover, we have the following properties:*

(1) If $\mu(A) = 0$, $A \subseteq X$, then $A \in \mathcal{M}_\mu$ (i.e., every $\mu-$negligible set is $\mu-$measurable).

(2) μ is $\sigma-$additive on $\mathcal{M}_\mu$.

(3) If $\{A_i\} \subseteq \mathcal{M}_\mu$, then

$$(3.a) \quad \mu\left(\bigcup_{i=1}^{\infty} A_i\right) = \lim_{i \to \infty} \mu(A_i) \text{ provided } A_1 \subseteq A_2 \subseteq \ldots.$$

$$(3.b) \quad \mu\left(\bigcap_{i=1}^{\infty} A_i\right) = \lim_{i \to \infty} \mu(A_i) \text{ provided } A_1 \supseteq A_2 \supseteq \ldots \text{ and } \mu(A_1) < \infty.$$

In particular, given a measure μ on X we can always find a $\sigma-$algebra $\mathcal{M}_\mu$ restricted to which μ is $\sigma-$additive. Reciprocally, given a $\sigma-$algebra Σ and a $\sigma-$additive measure $\mu : \Sigma \to [0, \infty]$, we can extend μ to the whole power set of X as follows: for each $A \subseteq X$, define

$$\mu(A) := \inf\left\{\sum_i \mu(S_i) : \{S_i\}_{i \geqslant 1} \subseteq \Sigma \text{ with } A \subseteq \bigcup_i S_i\right\}.$$

It is straightforward to check this indeed defines a measure on X whose restriction to Σ is the originally given measure.

Definition A.4 ($\mu-$measurable functions)**.** Let Y be a topological space and μ a measure on X. A function $f : X \to Y$ is said to be $\mu-$**measurable** when $f^{-1}(U)$ is a $\mu-$measurable set in X for every open subset $U \subseteq Y$.

Given a measure μ on X, we do integration theory with respect to μ by restricting ourselves to the natural measure space $(X, \mathcal{M}_\mu, \mu|_{\mathcal{M}_\mu})$ determined by μ. The reader interested in the details of Lebesgue integration theory on measure spaces is kindly referred to Folland (ibid.).

Definition A.5. Let μ be a measure on X; we say that μ is:

- **locally finite** if $\mu(K) < \infty$ for each compact subset $K \subseteq X$;

- **metric** if $\mu(A \cup B) = \mu(A) + \mu(B)$, for each $A, B \subseteq X$ such that $d(A, B) > 0$;

- **Borel** if all Borel sets are μ–measurable, i.e. $\mathcal{B}(X) \subseteq \mathcal{M}_\mu$;

- **Borel regular** if it is a Borel measure and if for every $A \subseteq X$ there is a Borel set $B \subseteq X$ such that $A \subseteq B$ and $\mu(A) = \mu(B)$;

- **Radon** if it is a locally finite and Borel regular measure.

Let μ be a measure on X. If $f : X \to \mathbb{R}$ is a nonnegative ($f \geqslant 0$) μ–measurable function, then we can form a new measure $f\mu$ on X such that

$$[f\mu](A) := \int_A f \, \mathrm{d}\mu, \quad \forall A \in \mathcal{M}_\mu.$$

In particular, when $A \subseteq X$ is a μ–measurable subset, we denote by $\mu\lfloor A$ the measure $\chi_A \mu$, i.e.

$$[\mu\lfloor A](E) := \mu(A \cap E), \quad \forall E \subseteq X.$$

As one may check directly, all μ–measurable sets are $(\mu\lfloor A)$–measurable sets. Also, if μ is a Borel regular measure, then so is $\mu\lfloor A$. Moreover, it is not difficult to show the following:

Lemma A.6. *If μ is a Radon measure and $f \in L^1(\mu)$ is a nonnegative function, then $f\mu$ is a Radon measure.*

A key tool to check Borel sets are μ–measurable is the following Simon (1983, p. 3, Theorem 1.2):

Theorem A.7 (Carathéodory's criterion). *Let μ be a measure on the metric space X. Then,*

$$\mu \text{ is a Borel measure} \quad \Longleftrightarrow \quad \mu \text{ is a metric measure.}$$

Finally, we state a very useful approximation result for Borel regular measures Simon (ibid., Theorem 1.3 and Remark 1.4):

Theorem A.8 (Inner and outer approximation). *Suppose μ is a Borel regular measure on X and $X = \cup_{j \geqslant 1} V_j$, where $\mu(V_j) < \infty$ and V_j is open for each $j \in \mathbb{N}$. Then:*

(i) $\mu(A) = \inf\{\mu(U) : U \supseteq A, \ U \text{ open}\}$, for any $A \subseteq X$.

(ii) $\mu(A) = \sup\{\mu(C) : C \subseteq A,\ C\ \text{closed}\}$, *for any* $A \in \mathcal{M}_\mu$.

In particular, if X is a second countable and locally compact metric space[1] (e.g. when X is a manifold) and μ is a Radon measure on X then (i) holds and (ii) can be improved to

(ii') $\mu(A) = \sup\{\mu(K) : K \subseteq A,\ K\ \text{compact}\}$, *for any* $A \in \mathcal{M}_\mu$.

A.2 Hausdorff measure and dimension

We start describing a standard general process for constructing metric measures on metric spaces, called **Carathéodory's construction**. The input for this method is the following. Suppose we are given a pair $(\mathcal{F}, \rho)$, where $\mathcal{F}$ is a collection of subsets of X, $\rho \colon \mathcal{F} \to [0, \infty]$ and

(i) for each $\delta > 0$, there exists a countable cover $\{E_i\}_{i \in \mathbb{N}} \subseteq \mathcal{F}$ of X such that $\mathrm{diam}(E_i) \leqslant \delta$.

(ii) for each $\delta > 0$, there exists an element $E \in \mathcal{F}$ such that $\rho(E) \leqslant \delta$ and $\mathrm{diam}(E) \leqslant \delta$.

For example, if $\mathcal{F}$ contains all non-empty open balls of X and X is separable, then (i) is easily seen to be verified. If, moreover, one has $\rho(\cdot) = C\,\mathrm{diam}(\cdot)$, for some uniform constant $C \leqslant 1$, then (ii) is also checked trivially.

For each $\delta > 0$, define

$$\mathcal{F}_\delta := \{E \in \mathcal{F} : \mathrm{diam}(E) \leqslant \delta\},$$

and construct preliminary measures ν_δ on X putting, for each $A \subseteq X$,

$$\nu_\delta(A) := \inf\left\{\sum_{i=1}^{\infty} \rho(E_i) : A \subseteq \bigcup_{i=1}^{\infty} E_i \text{ and } \{E_i\}_{i \in \mathbb{N}} \subseteq \mathcal{F}_\delta\right\}.$$

We note that $0 < \delta \leqslant \delta'$ implies $\mathcal{F}_\delta \subseteq \mathcal{F}_{\delta'}$, so that $\nu_\delta \geqslant \nu_{\delta'}$. Therefore, it is well-defined (possibly ∞)

$$\nu(A) := \lim_{\delta \searrow 0} \nu_\delta(A) = \sup_{\delta > 0} \nu_\delta(A), \quad \text{for each } A \subseteq X.$$

---

[1] In particular, X admits an open covering $\{V_j\}$ such that $\overline{V}_j$ is compact and contained in V_{j+1}, for each $j \in \mathbb{N}$; see e.g. Warner (2013, proof of Lemma 1.9, p. 9).

It is straightforward to check that ν is a measure on X. Moreover, we claim that ν is in fact a metric measure (therefore, by Theorem (A.7), ν is Borel): indeed, if $A, B \subseteq X$ are such that $d(A, B) > \delta > 0$, then

$$\nu_\delta(A \cup B) \geq \nu_\delta(A) + \nu_\delta(B),$$

because whenever $\mathcal{C} = \{E_i\}$ is a covering of $A \cup B$ with $\operatorname{diam}(E_i) < \delta$, the collections

$$\mathcal{C} \cap \{E : E \cap A \neq \emptyset\} \text{ and } \mathcal{C} \cap \{E : E \cap B \neq \emptyset\}$$

are clearly disjoint, covering A and B respectively. Thus the claim follows from the definition of ν.

Example A.9 (Lebesgue measure). Let $X = \mathbb{R}^n$ and d be the usual Euclidean distance:

$$d(x, y) = \left(\sum_{i=1}^{n} (y_i - x_i)^2 \right)^{1/2}, \quad \text{for each } x, y \in \mathbb{R}^n.$$

Define

$$\mathcal{F} := \left\{ \prod_{i=1}^{n} [a_i, b_i] : a_i, b_i \in \mathbb{R}, a_i < b_i, i = 1, \ldots, n \right\},$$

i.e. $\mathcal{F}$ is the collection of all (non-degenerated) closed $n-$cubes on $\mathbb{R}^n$, and take $\rho \colon \mathcal{F} \to [0, \infty]$ defined by

$$\rho(\prod_{i=1}^{n} [a_i, b_i]) := \prod_{i=1}^{n} (b_i - a_i).$$

Then, the resulting measure of Carathéodory's construction applied to $(\mathcal{F}, \rho)$ is the $n-$**dimensional Lebesgue measure** $\mathcal{L}^n$ on $\mathbb{R}^n$.

A well-known characterization for $\mathcal{L}^n$ is the following: $\mathcal{L}^n$ is the unique Borel regular, translation-invariant measure on $\mathbb{R}^n$, normalized so that the measure of the unit cube $[0, 1]^n$ is 1 (see Federer (1969) and Simon (1983, p. 8)).

We now proceed to define the Hausdorff $s-$dimensional measure $\mathcal{H}^s$ on an arbitrary *separable* metric space (X, d).

Definition A.10 (Hausdorff measure). Let $s \in \mathbb{R}_{\geq 0}$. The $s-$**dimensional Hausdorff measure** $\mathcal{H}^s$ of a separable metric space (X, d) is the measure on X

generated by Carathéodory's construction when $\mathcal{F}$ is taken to be the collection of all non-empty subsets of X and ρ is given by

$$\rho(A) := 2^{-s}\operatorname{diam}(A)^s, \quad \text{for each } \emptyset \neq A \subseteq X.$$

More explicitly, for each $A \subseteq X$,

$$\mathcal{H}^s(A) := \lim_{\delta \downarrow 0} \mathcal{H}^s_\delta(A),$$

where, for each $\delta > 0$,

$$\mathcal{H}^s_\delta(A) := 2^{-s} \inf\left\{ \sum_{i=1}^{\infty} (\operatorname{diam}(E_i))^s : A \subseteq \bigcup_{i=1}^{\infty} E_i, \text{ and } \operatorname{diam}(E_i) \leqslant \delta \right\}.$$

Example A.11 (n–dimensional Hausdorff measure of a connected Riemannian n–manifold). Let (M, g) be a connected Riemannian n–manifold. Then, on the one hand, viewing M as a metric space with the natural Riemannian distance function d_g induced by g (see e.g. Aubin (1982, §2.1)), we get associated s–dimensional Hausdorff measures $\mathcal{H}^s$ on M for each nonnegative real number s; in particular, we get $\mathcal{H}^n$.

On the other hand, supposing further that M is oriented, we get a Riemannian volume n–form $\mathrm{d}V_g$ on (M, g), which in turn induces a canonical Radon measure μ_g on M resulting from the application of Riesz's representation theorem (see Remark A.26) on the integration functional

$$I_g : C_c(M; \mathbb{R}) \to \mathbb{R}$$

$$f \mapsto \int_M f \, \mathrm{d}V_g.$$

Now, for all $s \in \mathbb{R}_{\geqslant 0}$, we define

$$\alpha_s := \frac{\pi^{\frac{s}{2}}}{\Gamma\left(\frac{s}{2} + 1\right)},$$

where $\Gamma(z) := \int_0^\infty x^{z-1} e^{-x} \, dx$ (defined for each $z \in \mathbb{C}$ with positive real part) is the so-called *Euler gamma function*. Note that when $s = k \in \mathbb{N}_0$ is a nonnegative integer, α_k is precisely the Lebesgue measure $\mathcal{L}^k(B_1(0))$ of the unit ball in $\mathbb{R}^k$.

In this setting, we can state the following relation between $\mathcal{H}^n$ and μ_g:

Proposition A.12. *On a connected, oriented, Riemannian n-manifold (M, g), the n-dimensional Hausdorff measure $\mathcal{H}^n$ multiplied by the constant factor α_n equals the Riemannian volume measure μ_g.*

The reader interested in a proof of this fact may consult D. Burago, Y. Burago, and Ivanov (2001, p. 196, Theorem 5.5.5) (see also Simon (1983, p. 10, Theorem 2.6) for the $M = \mathbb{R}^n$ case).

The following proposition is immediate from the above definitions:

Proposition A.13. *Let $A \subseteq X$ and $s \in \mathbb{R}_{\geq 0}$. Then:*

(i) $\mathcal{H}^s(A) < \infty \Rightarrow \mathcal{H}^{s'}(A) = 0,\ \forall s' > s$.

(ii) $\mathcal{H}^s(A) > 0 \Rightarrow \mathcal{H}^{s''}(A) = \infty,\ \forall s'' < s$.

We can then make the following definition:

Definition A.14 (Hausdorff dimension). The **Hausdorff dimension** $\dim_{\mathcal{H}}(A)$ of a subset $A \subseteq X$ is the extended real number given by

$$\dim_{\mathcal{H}}(A) := \inf\{s \in \mathbb{R}_{\geq 0} : \mathcal{H}^s(A) = 0\} = \sup\{s \in \mathbb{R}_{\geq 0} : \mathcal{H}^s(A) = \infty\}.$$

In other words, if $A \subseteq X$ then the Hausdorff dimension $\dim_{\mathcal{H}}(A)$ of A is the unique extended real number in $[0, \infty]$ such that

$$\begin{aligned}
s < \dim_{\mathcal{H}}(A) \quad &\Rightarrow \quad \mathcal{H}^s(A) = \infty, \\
s > \dim_{\mathcal{H}}(A) \quad &\Rightarrow \quad \mathcal{H}^s(A) = 0.
\end{aligned}$$

A priori, in case $s = \dim_{\mathcal{H}}(A)$, all the three possibilities $\mathcal{H}^s(A) = 0$, $0 < \mathcal{H}^s(A) < \infty$ and $\mathcal{H}^s(A) = \infty$ are admissible. On the other hand, if we can find s such that $0 < \mathcal{H}^s(A) < \infty$, then certainly $\dim_{\mathcal{H}}(A) = s$. Also, if $s \in \mathbb{R}_{\geq 0}$ is such that $\mathcal{H}^s(A) < \infty$ then $\dim_{\mathcal{H}}(A) \leq s$.

Some immediate properties the Hausdorff dimension satisfies are the following:

- (Monotonicity) If $A \subseteq B \subseteq X$, then $\dim_{\mathcal{H}}(A) \leq \dim_{\mathcal{H}}(B)$;

- (Stability w.r.t. countable unions) If $\{A_i\}$ is a countable collection of subsets $A_i \subseteq X$, then

$$\dim_{\mathcal{H}}\left(\bigcup_i A_i\right) = \sup_i \dim_{\mathcal{H}}(A_i).$$

In particular, if $S \subseteq X$, is such that $S = \bigcup_{i \geq 1} A_i$ with $\mathcal{H}^s(A_i) < \infty$ (for each $i \geq 1$), then $\dim_{\mathcal{H}}(S) \leq s$.

A.3 Densities and covering theorems

We start this section summarizing the covering theorems that are particularly useful for this work and then introduce the notion(s) of (lower and upper) density(ies) of measures. We finish with results relating appropriate information about the upper density of a measure and relations between such measure and the Hausdorff measure, as well as estimates on the upper density of the Hausdorff measure on appropriate sets.

Covering theorems. The first lemma we shall prove is a simple but useful result on metric space topology.

Lemma A.15. *Let K be a compact subspace of a metric space (X, d). Given $r > 0$, we can find a finite set of points $\{x_1, \ldots, x_m\} \subseteq K$ such that the following holds:*

(i) $K \subseteq \bigcup_{i=1}^{m} B_{2r}(x_i)$, and

(ii) $B_r(x_i) \cap B_r(x_j) = \emptyset$, for each $i \neq j$, $i, j \in \{1, \ldots, m\}$.

Proof. We describe an explicit algorithm to construct the $\{x_i\}$. In the first step, fix some $x_1 \in K$. In the second step, consider

$$C_2 := K \setminus B_{2r}(x_1).$$

If $C_2 = \emptyset$, stop the algorithm; the set $\{x_1\}$ will do the job. If $C_2 \neq \emptyset$, then choose $x_2 \in C_2$ and go to the next step. In general, when we arrive at the j–th step, $j \geqslant 2$, the first $j - 1$ points $x_1, \ldots, x_{j-1} \in K$ are already constructed, so we consider

$$C_j := K \setminus \bigcup_{i=1}^{j-1} B_{2r}(x_i).$$

If $C_j = \emptyset$, stop the algorithm; the set $\{x_1, \ldots, x_{j-1}\}$ is clearly the set of points in K we are looking for. If $C_j \neq \emptyset$, choose $x_j \in C_j$ and go to the next step.

   We claim this process ends in a finite number of steps, i.e. we always arrive at the case $C_j = \emptyset$, for some $j \in \mathbb{N}$ large enough. Otherwise, the algorithm just described would give rise to a sequence $\{x_i\}_{n=1}^{\infty}$ in K which does not admit a convergent subsequence: if $n, m \in \mathbb{N}$ are such that $n < m$, then $d(x_n, x_m) > 2r$ because $x_m \notin \bigcup_{i=1}^{m-1} B_{2r}(x_i) \supseteq B_{2r}(x_n)$ by construction. This contradicts the compacity of K. $\qquad\square$

Another important covering theorem is the following (cf. Simon (1983, Theorem 3.3)).

Theorem A.16 (5r−covering lemma)**.** *Suppose* (X, d) *is a* separable *metric space. If* $\mathscr{B}$ *is an arbitrary family of (closed or open) balls in* X *satisfying*

$$\sup_{B \in \mathscr{B}} \, \mathrm{diam}(B) < \infty,$$

then there exists a countable *and (pairwise) disjoint* subcollection $\mathscr{B}' \subseteq \mathscr{B}$ *such that*

$$\bigcup_{B \in \mathscr{B}} B \subseteq \bigcup_{B \in \mathscr{B}'} 5B,$$

where $5B$ *denotes a ball with the same center as* B *and five times the radius of* B.

Densities.   We now introduce the notions of upper and lower k−dimensional density of a measure at a point. The reference for this part is Simon (ibid., §3).

Definition A.17 (Upper and lower densities). Let $s \in \mathbb{R}_{\geq 0}$ and let μ be a measure on X. We define the **upper** (resp. **lower**) s−**dimensional density** of μ at $x \in X$ by

$$\Theta^{*s}(\mu, x) := \limsup_{r \downarrow 0} r^{-s} \mu(B_r(x)).$$

$$\left(\text{resp. } \Theta_*{}^s(\mu, x) := \liminf_{r \downarrow 0} r^{-s} \mu(B_r(x)). \right)$$

Whenever $\Theta^{*s}(\mu, x) = \Theta_*{}^s(\mu, x)$, we denote the common value by

$$\Theta(\mu, x) := \lim_{r \downarrow 0} r^{-s} \mu(B_r(x))$$

and simply speak of the s−**density of** μ **at** x.

For an arbitrary subset $A \subseteq X$, we define the **upper** (resp. **lower**) s−**dimensional density** of A at x by

$$\Theta^{*s}(A, x) := \Theta^{*s}(\mathcal{H}^s \lfloor A, x).$$

$$(\text{resp. } \Theta_*{}^s(A, x) := \Theta_*{}^s(\mathcal{H}^s \lfloor A, x).)$$

When the upper and lower s−dimensional densities of A at x are equal we write the common value by $\Theta^s(A, x)$.

Remark A.18. Some authors (including L. Simon) define the Hausdorff measure multiplying the one in Definition (A.10) by the constant factor α_s of Example A.11. In this case, it is convenient to modify the above definition multiplying the densities by α_s^{-1}. $\diamond$

Remark A.19. We note that when μ is a Borel measure then $\Theta^{*s}(\mu, \cdot)$ and $\Theta_*^{\ s}(\mu, \cdot)$ are μ–measurable functions. In fact, for each fixed $r > 0$, the function on X defined by

$$x \mapsto \mu(B_r(x))$$

is upper semi-continuous whenever μ is a Borel measure. Indeed: fix $x \in X$ and $r > 0$; we want to show

$$\mu(B_r(x)) \geq \limsup_{y \to x} \mu(B_r(y)).$$

If $\mu(B_r(x)) = \infty$ the assertion is clearly true, so suppose $\mu(B_r(x)) < \infty$. Let (x_n) be a sequence in X such that $x_n \to x$. Then, given $\varepsilon > 0$, there exists $n_0 \in \mathbb{N}$ such that

$$n \geq n_0 \quad \Rightarrow \quad B_r(x_n) \subseteq B_{r+\varepsilon}(x).$$

Thus, on one hand we have

$$\limsup_{n \to \infty} \mu(B_r(x_n)) \leq \mu(B_{r+\varepsilon}(x)).$$

On the other hand, for $\varepsilon_0 > 0$ small enough, Theorem A.3 (3.b) implies

$$\mu(B_r(x)) = \mu\left(\bigcap_{0 < \varepsilon < \varepsilon_0} B_{r+\varepsilon}(x)\right) = \lim_{\varepsilon \downarrow 0} \mu(B_{r+\varepsilon}(x)).$$

Therefore

$$\limsup_{n \to \infty} \mu(B_r(x_n)) \leq \mu(B_r(x)).$$

The claim follows. $\diamond$

Remark A.20. If $x \notin \operatorname{supp}(\mu)$ then $\Theta^s(\mu, x) = 0$ for every $0 \leq s < \infty$. Indeed, when $x \notin \operatorname{supp}(\mu)$ there exists an open subset $U \subseteq X$ such that $x \in U$ and $\mu(U) = 0$. Thus, for all sufficiently small $r > 0$ we have $\mu(B_r(x)) = 0$. In particular, $\Theta^s(\mu, x) = 0$ for every $0 \leq s < \infty$. $\diamond$

The next result tells us that appropriate information about the upper s–dimensional density function of a given Borel-regular measure gives estimates of this measure with respect to the s–dimensional Hausdorff measure.

Theorem A.21. *Let μ be a Borel regular measure on X, and let $s, t \in \mathbb{R}_{\geq 0}$.*

(i) If $A_1 \subseteq A_2 \subseteq X$ and $\Theta^{*s}(\mu \lfloor A_2, x) \geq t$ for all $x \in A_1$, then

$$t \mathcal{H}^s(A_1) \leq \mu(A_2).$$

(ii) If $A \subseteq X$ and $\Theta^{*s}(\mu \lfloor A, x) \leq t$ for all $x \in A$, then

$$\mu(A) \leq 2^s t \mathcal{H}^s(A).$$

In particular, (i) and (ii) imply

$$t_1 \mathcal{H}^s(A) \leq \mu(A) \leq 2^s t_2 \mathcal{H}^s(A),$$

whenever $A \subseteq X$ is such that $0 \leq t_1 \leq \Theta^{*s}(\mu \lfloor A, x) \leq t_2$ for all $x \in A$.

The proof of the above result uses Theorem A.16 for (i) and is elementary for (ii); see Simon (1983, Theorem 3.2). As a corollary of Theorem A.21 (i), one can prove the following useful result Simon (ibid., Theorem 3.5):

Theorem A.22. *If μ is Borel-regular and $A \subseteq X$ is μ−measurable with $\mu(A) < \infty$ then*

$$\Theta^{*s}(\mu \lfloor A, x) = 0 \quad \text{for } \mathcal{H}^s - a.e. \ x \in X \setminus A.$$

Restricting attention to Hausdorff measures, there are some useful estimates for the density on sets of finite measure Simon (ibid., Theorem 3.6).

Theorem A.23. *Let $s \in \mathbb{R}_{\geq 0}$. Then the following assertions holds.*

(i) If $\mathcal{H}^s(A) < \infty$ then $\Theta^{*s}(A, x) \leq 1$ for $\mathcal{H}^s$−a.e. $x \in A$.

(ii) If $\mathcal{H}^s_\delta(A) < \infty$ for each $\delta > 0$, then $\Theta_*{}^s(A, x) \geq 2^{-s}$ for $\mathcal{H}^s$−a.e. $x \in A$.

In particular[2], if $\mathcal{H}^s(A) < \infty$ then

$$2^{-s} \leq \Theta^{*s}(A, x) \leq 1 \quad \text{for } \mathcal{H}^s - a.e. \ x \in A.$$

---

[2]Note that $\mathcal{H}^s \geq \mathcal{H}^s_\delta \geq \mathcal{H}^s_\infty$.

A.4 Radon measures

In this subsection we assume X is a *locally compact* and *separable* metric space.

Let H denote a finite dimensional real *Hilbert space* with inner product $\langle \cdot, \cdot \rangle$ and induced norm $\| \cdot \|$. Denote by $C_c^0(X; H)$ the space of continuous functions $X \to H$ with *compact* support in X. We endow $C_c^0(X; H)$ with the topology of uniform convergence on compact sets: if $\{f_n\}_{n \in \mathbb{N}} \subset C_c^0(X; H)$, then $f_n \to f \in C_c^0(X; H)$ if, and only if,

1. there exists a compact subset $K \subseteq X$ such that $\operatorname{supp}(f_n) \subseteq K$, for each $n \in \mathbb{N}$; and

2. $\sup \{\| f_n(x) - f(x)\| : x \in K\} \to 0$ as $n \to \infty$.

Given a Radon measure μ on X and a μ–measurable function $v \colon X \to H$ with $\|v(x)\| = 1$ for μ-a.e. $x \in X$, then

$$L \colon f \mapsto \int_X \langle f, v \rangle d\mu$$

defines a continuous linear functional on $C_c^0(X; H)$: indeed, let $K \subseteq X$ be a compact set and suppose $f \in C_c^0(X; H)$ is such that $\operatorname{supp}(f) \subseteq K$. Since f is continuous and μ is Radon, we have $\| f \|_\infty = \sup_{x \in K} \| f(x) \| < \infty$ and $C_K := \mu(K) < \infty$. Moreover, by the hypothesis on v and the Cauchy–Schwarz inequality,

$$|\langle f(x), v(x) \rangle| \leqslant \| f(x) \| \| v(x) \| = \| f(x) \|, \quad \text{for } \mu\text{-a.e. } x \in X.$$

Therefore,

$$|L(f)| \leqslant \int_K |\langle f(x), v(x) \rangle| d\mu(x) \leqslant C_K \| f \|_\infty.$$

Conversely, we have Simon (ibid., Theorem 4.1, p.18):

Theorem A.24. *(Riesz) Let $L \colon C_c^0(X; H) \to \mathbb{R}$ be a linear functional such that*

$$\sup\{L(f) : f \in C_c^0(X; H), \ \| f \|_\infty \leqslant 1, \ \operatorname{supp}(f) \subseteq K\} < \infty, \quad \text{(A.25)}$$

for each compact $K \subseteq X$. Then there is a Radon measure μ on X and a μ–measurable function $v \colon X \to H$ with $\|v(x)\| = 1$ for μ-a.e. $x \in X$ such that

$$L(f) = \int_X \langle f, v \rangle \mathrm{d}\mu, \quad \forall f \in C_c^0(X; H).$$

Moreover, the Radon measure μ is unique; in fact, in the above conditions we have

$$\mu(V) = \sup\{L(f) : f \in C_c^0(X; H),\ \|f\|_\infty \leq 1\ \text{and}\ \text{supp}(f) \subseteq V\},$$

for every open subset $V \subseteq X$; μ is called the **total variation measure** *associated with the functional L.*

Remark A.26. When $H = \mathbb{R}$, if we replace the hypothesis (A.25) in Theorem A.24 by the condition that $Lf \geq 0$ whenever $f \geq 0$ (in case L is called a *nonnegative* functional), then we can in fact find $v \colon X \to \mathbb{R}$ such that $v \equiv 1$ μ−a.e. and, therefore, conclude that

$$L(f) = \int_X f\, d\mu, \quad \forall f \in C_c^0(X; \mathbb{R}).$$

Such version of the Riesz representation theorem can be found, for example, in Folland's book Folland (2013, Theorem 7.2, p.212) (see also Rudin (1986, Theorem 2.14, p.40)). In particular, we can identify the set of Radon measures on X with the set of nonnegative linear functionals on $C_c^0(X) := C_c^0(X; \mathbb{R})$.

$$\Diamond$$

It is then natural to endow the space of Radon measures on X with the weak* topology of the topological dual of $C_c^0(X)$:

Definition A.27 (Weak* convergence). Given a sequence of Radon measures $\{\mu_i\}$ we say that μ_i **converges weakly*** to a Radon measure μ, and we write $\mu_i \rightharpoonup \mu$, when

$$\lim_{i \to \infty} \int_X f\, d\mu_i = \int_X f\, d\mu, \quad \forall f \in C_c^0(X).$$

Having Remark A.26 in mind, the following theorem is a fairly easy application of the general Banach–Alaoglu theorem.

Theorem A.28 (Weak* Compactness of Radon Measures). *If $\{\mu_i\}$ is a sequence of Radon measures on X satisfying*

$$\sup\{\mu_i(U) : i \geq 1\} < \infty, \quad \forall U \Subset X,$$

then $\{\mu_i\}$ admits a weakly convergent subsequence.*

The following basic result is of fundamental importance and will be used repeatedly in Chapter 4 De Lellis (2008, Proposition 2.7, p.8).

Theorem A.29. *Let $\{\mu_i\}$ be a sequence of Radon measures on X such that $\mu_i \rightharpoonup \mu$.*

(i) If $U \subseteq X$ is an open subset then

$$\mu(U) \leqslant \liminf_{i \to \infty} \mu_i(U).$$

(ii) If $K \subseteq X$ is a compact subset then

$$\mu(K) \geqslant \limsup_{i \to \infty} \mu_i(K).$$

In particular,

(iii) If $U \Subset X$ is a precompact open subset with $\mu(\partial U) = 0$, then

$$\mu(U) = \lim_{i \to \infty} \mu_i(U).$$

(iv) Given $x \in X$ and $\delta > 0$, then

$$\mathscr{R}_{x,\delta}(\mu) := \{r \in \,]0, \delta] : \mu(\partial B_r(x)) > 0\}$$

is at most countable and

$$\mu(B_r(x)) = \lim_{i \to \infty} \mu_i(B_r(x)), \quad \forall r \in \,]0, \delta] \setminus \mathscr{R}_{x,\delta}(\mu).$$

We end this section with a theorem which requires the following definitions.

Definition A.30. Let $\mathscr{B}$ be a collection of balls in X. We define the **set of centres** of $\mathscr{B}$ to be

$$C_{\mathscr{B}} := \{x \in X : B_r(x) \in \mathscr{B} \text{ for some } r > 0\}.$$

A subset $A \subseteq X$ is said to be covered **finely** by $\mathscr{B}$ if for every $x \in A$ and every $\varepsilon > 0$ there exists a ball $B \in \mathscr{B}$ such that $x \in B$ and $\mathrm{diam}(B) < \varepsilon$.

Definition A.31. Let μ be a Radon measure on X. We say that X has the **symmetric Vitali property** relative to μ if for every collection of balls $\mathscr{B}$ which covers its set of centres $C_{\mathscr{B}}$ finely and with $\mu(C_{\mathscr{B}}) < \infty$, there is a countable pairwise disjoint subcollection $\mathscr{B}' \subseteq \mathscr{B}$ covering $\mu-$almost all of $C_{\mathscr{B}}$.

Example A.32. If X is locally compact, Hausdorff and second countable (e.g. if X is a manifold) then X has the symmetric Vitali property relative to every Radon measure on X.

The following is a useful result about differentiation of measures due to Besicovitch Simon (1983, p. 24, Theorem 4.7).

Theorem A.33 (Besicovitch differentiation of measures). *Suppose μ_1 and μ_2 are Radon measures on X, where X has the symmetric Vitali property with respect to μ_1. Then*

$$\frac{d\mu_2}{d\mu_1}(x) := \lim_{r \downarrow 0} \frac{\mu_2(B_r(x))}{\mu_1(B_r(x))}$$

exists (possibly ∞) μ_1–almost everywhere and defines a μ_1–measurable function on X. Furthermore, the Radon–Nikodým decomposition of μ_2 with respect to μ_1 is given by

$$\mu_2 = \frac{d\mu_2}{d\mu_1}\mu_1 + \mu_2 \lfloor Z, \tag{A.34}$$

where Z is a Borel set of μ_1–measure zero. Moreover, in case X also has the symmetric Vitali property with respect to μ_2 then $\dfrac{d\mu_2}{d\mu_1}$ also exists μ_2–almost everywhere and we may take $Z = \left\{ x : \dfrac{d\mu_2}{d\mu_1}(x) = \infty \right\}$ in (A.34).

A.5   Rectifiable sets and measures

Perhaps the most relevant class of functions in the context of geometric measure theory is the class of Lipschitz functions.

Definition A.35 (Lipschitz maps). A map $f : (X, d) \to (X', d')$ between metric spaces is called λ–**Lipschitz**, for some $\lambda \in [0, \infty[$, when

$$d'(f(x), f(y)) \leqslant \lambda d(x, y), \quad \forall x, y \in X.$$

Whenever

$$\mathrm{Lip}(f) := \inf\{\lambda \in [0, \infty[: f \text{ is } \lambda - \mathrm{Lipschitz}\} < \infty,$$

f is called a **Lipschitz** function.

Lemma A.36. *Let X and X' be metric spaces and $E \subseteq X$ an arbitrary subset. If $f : E \to X'$ is a Lipschitz map, then*

$$\mathcal{H}^s(f(E)) \leqslant \mathrm{Lip}(f)^s \mathcal{H}^s(E).$$

In particular, a Lipschitz map takes $\mathcal{H}^s$–negligible sets to $\mathcal{H}^s$–negligible sets.

Next, we give a simple extension result.

Lemma A.37. *Let $A \subseteq X$ and $n \in \mathbb{N}$. Then, every λ–Lipschitz map admits a $\sqrt{n}\lambda$–Lipschitz extension $\overline{f} : X \to \mathbb{R}^n$.*

Sketch of proof. For $n = 1$, we simply define

$$\overline{f}(x) := \inf\{f(a) + \lambda f(x) : a \in A\}, \quad \forall x \in X.$$

It is straightforward to verify $\overline{f}$ is well-defined and satisfy the desired properties.

For $n \geq 2$ one writes $f = (f_1, \ldots, f_n)$ and extends each $f_i : A \to \mathbb{R}$ separately as above. $\square$

For Lipschitz maps between Euclidean spaces we have the following important result Simon (ibid., Theorem 5.2, p.30).

Theorem A.38 (Rademacher). *If $f : \mathbb{R}^n \to \mathbb{R}^m$ is a Lipschitz map, then f is differentiable for $\mathcal{L}^n$–a.e. $x \in \mathbb{R}^n$.*

Using the above theorem (and other results), one can prove Simon (ibid., Theorem 5.3, p.32):

Theorem A.39. *If $U \subseteq \mathbb{R}^n$ is open and if $f : U \to \mathbb{R}$ is differentiable $\mathcal{L}^n$–a.e. in U, then for each $\varepsilon > 0$ there is a closed set $A \subseteq U$ and a C^1–function $g : \mathbb{R}^n \to \mathbb{R}$ such that*

$$\mathcal{L}^n(U \setminus A) < \varepsilon, \ f|_A = g|_A \ and \ (\mathrm{grad}\ f)|_A = (\mathrm{grad}\ g)|_A.$$

We now introduce a concept of great importance in geometric measure theory, which can be seen as a measure-theoretic notion of smoothness.

Definition A.40 (Rectifiable sets and measures). Let $k \in \mathbb{N}_0$. A subset $\Gamma \subseteq X$ is called **countably $\mathcal{H}^k$–rectifiable** if there exists a sequence of Lipschitz maps $f_i : A_i \subseteq \mathbb{R}^k \to X$ such that

$$\mathcal{H}^k\left(\Gamma \setminus \bigcup_i f_i(A_i)\right) = 0.$$

A Radon measure ν on X is called $\mathcal{H}^k$–**rectifiable** if $\nu = \theta \mathcal{H}^k \lfloor \Gamma$ for some countably $\mathcal{H}^k$–rectifiable set Γ and some Borel function $\theta : \Gamma \to [0, \infty[$.

By Lemma A.36, it follows that if $\Gamma \subseteq X$ is a countably $\mathcal{H}^k$–rectifiable set then $\mathcal{H}^k \lfloor \Gamma$ is σ–finite and, therefore, $\dim_{\mathcal{H}} \Gamma \leq k$. Note also that any Borel subset of a countably $\mathcal{H}^k$–rectifiable set is countably $\mathcal{H}^k$–rectifiable. Moreover, a countable union of countably $\mathcal{H}^k$–rectifiable sets is again a countably $\mathcal{H}^k$–rectifiable set.

Remark A.41. By definition, the property of being countably $\mathcal{H}^k$–rectifiable is *intrinsic*, i.e. if (X, d) is isometrically embedded in another metric space (X', d'), then $\Gamma \subseteq X$ is countably $\mathcal{H}^k$–rectifiable in X if, and only if, Γ is countably $\mathcal{H}^k$–rectifiable in X'. $\Diamond$

Here is a slightly different characterization of rectifiable sets that uses as A_i compact sets and that shows that the collection $\{f_i(A_i)\}$ can be disjoint Lang (2007, Proposition 9.2, p.20).

Theorem A.42. *Suppose X is a locally complete metric space and $\Gamma \subseteq X$ an $\mathcal{H}^k$–measurable and countably $\mathcal{H}^k$–rectifiable set. Then there exists a countable family of bi-Lipschitz[3] maps $f_i \colon K_i \to f(K_i) \subseteq \Gamma$, with $K_i \subseteq \mathbb{R}^k$ compact, such that the images $f_i(K_i)$ are pairwise disjoint and*

$$\mathcal{H}^k \left(\Gamma \setminus \bigcup_i f_i(K_i) \right) = 0.$$

For subsets of Euclidean spaces, using Theorem A.39, one has the following characterization of countably $\mathcal{H}^k$–rectifiable sets Simon (1983, Lemma 11.1, p.59).

Theorem A.43. *Let $E \subseteq \mathbb{R}^n$, where $n \geq k$. Then, E is countably $\mathcal{H}^k$–rectifiable if, and only if, there exists a sequence of k–dimensional C^1–submanifolds N_i of $\mathbb{R}^n$ such that*

$$\mathcal{H}^k \left(E \setminus \bigcup_i N_i \right) = 0.$$

More generally, in arbitrary complete Riemannian manifolds, one has important characterizations of rectifiability for both sets and measures in terms of *approximate tangent spaces*. In what follows, we will give a brief account of this topic. For a detailed discussion of the concept of rectifiability and its characterizations in Euclidean spaces the reader is encouraged to see DeLellis' lecture notes De Lellis (2008) (also see Simon's notes Simon (1983, p.60-66)). Here we will adapt the relevant definitions and results to the context of Riemannian manifolds.

---

[3]i.e. f_i is Lipschitz, injective and such that $f_i^{-1}|_{f_i(K_i)}$ is Lipschitz.

In what follows, let (M, g) be a connected, *complete*, Riemannian n−manifold. For each $s \in \mathbb{R}_{\geq 0}$, we let $\mathcal{H}^s$ denote the s−dimensional Hausdorff measure on M associated to the induced Riemannian distance function d_g.

Definition A.44 (s−tangent measures). Let ν be a Radon measure on M, and let $s \in \mathbb{R}_{\geq 0}$. Given $x \in M$ and $\lambda \in \mathbb{R}_+$, we write τ_λ for the linear scaling map on $T_x M$ taking v to λv, and define the scaled and translated measure $\nu_{x,\lambda} := (\exp_x \circ \tau_\lambda)^* \nu$ on $T_x M$ by

$$\nu_{x,\lambda}(E) = \nu(\exp_x(\lambda E)), \quad \forall E \subseteq T_x M.$$

We say that a Radon measure η on $T_x M$ is a s−**tangent measure of** ν **at** x when there exists a null-sequence $\{\lambda_i\} \subseteq \mathbb{R}_+$ such that

$$\lambda_i^{-s} \nu_{x,\lambda_i} \rightharpoonup \eta.$$

We let $\mathrm{Tan}_s(\nu, x)$ denote the set of all s−tangent measures of ν at x.

Remark A.45. Note that if $(M, g) = (\mathbb{R}^n, g_0)$, where g_0 is the standard Euclidean metric, then $\nu_{x,\lambda}(E) = \nu(x + \lambda E)$. $\diamond$

Theorem A.46 (Marstrand). *Let ν be a Radon measure on M, let $s \in \mathbb{R}_{\geq 0}$ and let $\Gamma \subseteq M$ be a Borel set with $\nu(\Gamma) > 0$. Suppose*

$$0 < \Theta_*^s(\nu, x) = \Theta^{*s}(\nu, x) < \infty \quad \text{for } \nu\text{−a.e. } x \in \Gamma.$$

Then $s = k \in \mathbb{N}_0$. Moreover, for ν−a.e. $x \in \Gamma$, there exists a k−dimensional subspace $V_x \leq T_x M$ such that $\Theta^k(\nu, x)\mathcal{H}^k \lfloor V_x \in \mathrm{Tan}_k(\nu, x)$.

From now on k will denote a nonnegative integer.

Definition A.47 (Approximate tangent spaces). Let $\Gamma \subseteq M$ be an $\mathcal{H}^k$−measurable set, and let $\Theta \colon \Gamma \to \;]0, \infty[$ be a locally $\mathcal{H}^k$−integrable function. A k−dimensional subspace $V_x \leq T_x M$ is called the **approximate** k−**tangent space** for Γ at x with multiplicity $\Theta(x)$ if $\mathrm{Tan}_k(\Theta\mathcal{H}^k \lfloor \Gamma, x) = \{\Theta(x)\mathcal{H}^k \lfloor V_x\}$, i.e. if

$$\lambda^{-k}(\Theta\mathcal{H}^k \lfloor \Gamma)_{x,\lambda} \rightharpoonup \Theta(x)\mathcal{H}^k \lfloor V_x \quad \text{as } \lambda \downarrow 0.$$

Let ν be a Radon measure on M. A k−dimensional subspace $V_x \leq T_x M$ is called the **approximate** k−**tangent space** for ν at x with multiplicity $\Theta(x) \in \;]0, \infty[$ if $\mathrm{Tan}_k(\nu, x) = \{\Theta(x)\mathcal{H}^k \lfloor V_x\}$, i.e. if

$$\lambda^{-k} \nu_{x,\lambda} \rightharpoonup \Theta(x)\mathcal{H}^k \lfloor V_x \quad \text{as } \lambda \downarrow 0.$$

Remark A.48. Let ν be a Radon measure on M and let $x \in M$. We claim that if $\eta := \Theta(x)\mathcal{H}^k \lfloor V_x \in \mathrm{Tan}_k(\nu, x)$ for some $\Theta(x) \in \,]0, \infty[$ and some k−dimensional subspace $V_x \leqslant T_x M$, then $\Theta^k(\nu, x)$ exists and equals $\Theta(x)$.

Pick $r \in \mathbb{R}_+$ such that $\eta(\partial B_r(0)) = 0$. Then

$$\Theta(x)r^k = \eta(B_r(0)) = \lim_{\lambda \downarrow 0} \lambda^{-k} \nu_{x,\lambda}(B_r(0))$$

$$= \lim_{\lambda \downarrow 0} \lambda^{-k} \nu(B_{\lambda r}(x)).$$

Therefore,

$$\Theta(x) = \lim_{\lambda \downarrow 0}(\lambda r)^{-k}\nu(B_{\lambda r}(x)) = \lim_{\delta \downarrow 0}\delta^{-k}\nu(B_\delta(x)),$$

from which our claim follows, since $\Theta(x) \in \,]0, \infty[$. $\diamondsuit$

We are now in position to state a number of rectifiability criteria.

Theorem A.49. *Let $\Gamma \subseteq M$ be an $\mathcal{H}^k$−measurable set. Then, Γ is countably $\mathcal{H}^k$−rectifiable if, and only if, there exists an $\mathcal{H}^k$−integrable function $\Theta\colon \Gamma \to \,]0, \infty[$ such that $\nu := \Theta\mathcal{H}^k \lfloor \Gamma$ has approximate k−tangent space V_x for $\mathcal{H}^k$−a.e. $x \in \Gamma$.*

Theorem A.50. *Let ν be a Radon measure on M. Then, ν is $\mathcal{H}^k$−rectifiable if, and only if, for ν−a.e. $x \in M$, there exist a positive constant $\Theta(x) \in \,]0, \infty[$ and a k−dimensional subspace $V_x \leqslant T_x M$ such that V_x is the approximate k−tangent space for ν at x with multiplicity $\Theta(x)$.*

The last result we cite is highly non-trivial and was proved in Preiss (1987) by David Preiss.

Theorem A.51 (Preiss). *Let ν be a locally finite Borel measure on M. Suppose that, for $k \in \mathbb{N}$, $k \leqslant n$,*

$$0 < \Theta_*^k(\nu, x) = \Theta^{k\,*}(\nu, x) < \infty, \quad \textit{for } \nu\textit{−a.e. } x \in \mathrm{supp}(\nu).$$

Then ν is $\mathcal{H}^k$−rectifiable.

A.6 Currents

In this section we introduce the basics about de Rham's theory of currents. Our goal is to establish the rectifiability Theorem A.68. We develop the theory in the framework of open subsets of $\mathbb{R}^n$ and at the end we explain how to pass from this context to arbitrary smooth manifolds.

The spaces $\Omega^k(U)$ and $\mathscr{D}^k(U)$. Let U be an open subset of $\mathbb{R}^n$. For each $\alpha = (\alpha_1, \ldots, \alpha_n) \in \mathbb{N}_0^n$ we associate the differential operator

$$D^\alpha := \left(\frac{\partial}{\partial x^1}\right)^{\alpha_1} \circ \ldots \circ \left(\frac{\partial}{\partial x^n}\right)^{\alpha_n},$$

whose *order* is

$$|\alpha| := \alpha_1 + \ldots + \alpha_n.$$

If $|\alpha| = 0$, then $D^\alpha = \mathbb{1}$.

As usual, we denote by $\Omega^k(U)$ the real vector space of *smooth k-forms* on U. We topologize $\Omega^k(U)$ with the C^∞_{loc}-**topology** which makes $\Omega^k(U)$ into a Fréchet space[4]. This is done by choosing an exhaustion of U by compact sets $\{K_i\}_{i \in \mathbb{N}}$ and defining, for each $i \in \mathbb{N}$, the semi-norm $p_i : \Omega^k(U) \to \mathbb{R}_{\geq 0}$ given by

$$p_i(\varphi) := \sup \left\{ \left|(D^\alpha \varphi_{j_1,\ldots,j_k})(x)\right| : x \in K_i,\ |\alpha| \leq i,\ 1 \leq j_1 < \ldots < j_k \leq n \right\},$$

for all

$$\varphi = \sum_{j_1 < \ldots < j_k} \varphi_{j_1,\ldots,j_k}\, \mathrm{d}x^{j_1 \cdots j_k} \in \Omega^k(U).$$

Then $\mathscr{P} := \{p_i\}$, being a countable separating family of semi-norms on $\Omega^k(U)$, defines a metrizable locally convex topology on $\Omega^k(U)$ admitting a translation-invariant compatible metric (see Rudin (1991, Theorem 1.37 and Remark (c) of Section 1.38)). A local base for 0 is given by the sets

$$V_i := \left\{ \varphi \in \Omega^k(U) : p_i(\varphi) < \frac{1}{i} \right\}, \quad i \in \mathbb{N}.$$

One may readily check that every Cauchy sequence on $\Omega^k(U)$ has a limit in $\Omega^k(U)$, so that $\Omega^k(U)$ is in fact a Fréchet space.

For each compact $K \subseteq U$,

$$\mathscr{D}^k_K(U) := \Omega^k(U) \cap \{\varphi : \operatorname{supp}(\varphi) \subseteq K\}$$

is a closed subspace of $\Omega^k(U)$, and therefore is also a Fréchet space. The union of the spaces $\mathscr{D}^k_K(U)$, as K ranges over all compact subsets of U, is denoted by

$$\mathscr{D}^k(U) := \Omega^k(U) \cap \{\varphi : \varphi \text{ has compact support in } U\}.$$

---

[4]i.e. a locally convex topological vector space whose topology is induced by a translation-invariant metric which makes the space complete.

This is clearly a vector space under the usual operations. We endow $\mathscr{D}^k(U)$ with the largest topology making the inclusion maps $\mathscr{D}^k_K(U) \hookrightarrow \mathscr{D}^k(U)$ continuous (cf. Federer (1978, §6)); this is called the C^∞-**topology** on $\mathscr{D}^k(U)$. It can be shown that this topology makes $\mathscr{D}^k(U)$ into a locally convex topological vector space. Moreover:

Proposition A.52. *Given $\{\varphi^i\}_{i \in \mathbb{N}} \subseteq \mathscr{D}^k(U)$, where*[5]

$$\varphi^i = \sum_J \varphi^i_J \, dx^J, \quad \text{for each } i \in \mathbb{N},$$

then $\varphi^i \to 0$ in $\mathscr{D}^k(U)$ if, and only if, the following holds:

   (i) there exists a compact subset $K \subseteq U$ with $\mathrm{supp}(\varphi^i) \subseteq K$, for all $i \in \mathbb{N}$.

   (ii) $\displaystyle\sup_{x \in K} |(D^\alpha \varphi^i_J)(x)| \to 0$ as $i \to \infty$, for all J and $\alpha \in \mathbb{N}_0^n$.

Proposition A.53. *Let $T : \mathscr{D}^k(U) \to Y$ be a linear map into a locally convex space Y. Then the following are equivalent:*

   (a) T is continuous.

   (b) If $\varphi^i \to 0$ in $\mathscr{D}^k(U)$ then $T\varphi^i \to 0$ in Y.

Remark A.54. The approach given above for the spaces $\Omega^k(U)$ and $\mathscr{D}^k(U)$ is an adaptation of Rudin's approach Rudin (1991, §1.46 and §6.2-6.8) for the corresponding spaces of functions. In this spirit, the reader interested in a proof of the above results may want to compare Proposition A.52 with Rudin (ibid., Theorem 6.5 (f), pp. 154-155), and Proposition A.53 with Rudin (ibid., Theorem 6.6 (a), (c), p. 155). $\Diamond$

An element $\varphi \in \mathscr{D}^k(U)$ is also known as a *test form*.

Definition A.55 (Current). A k-**current** T on U is an element of the topological dual $\mathscr{D}_k(U) := \left(\mathscr{D}^k(U)\right)'$, i.e. a continuous linear functional of $\mathscr{D}^k(U)$. When $k = 0$ we use the notations $\mathscr{D}(M) := \mathscr{D}^0(U)$ and $\mathscr{D}'(M) := \mathscr{D}_0(U)$.

0-currents are also known as *distributions*.

Definition A.56 (Weak* convergence). A sequence of k-currents $\{T_i\} \subseteq \mathscr{D}_k(U)$ **converges weakly*** to $T \in \mathscr{D}_k(U)$, and we write $T_i \rightharpoonup T$, if

$$\lim_{i \to \infty} T_i(\varphi) = T(\varphi), \quad \forall \varphi \in \mathscr{D}^k(U).$$

---

[5]Here the sum runs over all $J = \{1 \leqslant j_1 < \ldots < j_k \leqslant n\}$.

Let $T \in \mathscr{D}_k(M)$. The **support** $\mathrm{supp}(T)$ of T is the intersection of all closed subsets $F \subseteq M$ satisfying:

$$\mathrm{supp}(\varphi) \cap F = \emptyset, \ \varphi \in \mathscr{D}^k(M) \quad \Longrightarrow \quad T(\varphi) = 0.$$

Note that every compactly supported $k-$current extends to a continuous linear functional on $\Omega^k(U)$.

Definition A.57 (Boundary). Let $k \in \mathbb{N}$. If $T \in \mathscr{D}_k(U)$, the **boundary** of T is the current $\partial T \in \mathscr{D}_{k-1}(U)$ given by

$$\partial T(\varphi) := T(\mathrm{d}\varphi), \quad \forall \varphi \in \mathscr{D}^{k-1}(U).$$

We define the boundary of a $0-$current to be the zero function. A current $T \in \mathscr{D}_k(U)$ is said to be **closed** if $\partial T = 0$.

Remark A.58. We list some elementary observations concerning Definition A.57.

- $\partial \circ \partial = 0$, as a direct consequence of $\mathrm{d} \circ \mathrm{d} = 0$. In particular, there is an associated complex:

$$\ldots \to \mathscr{D}^{k+1}(U) \xrightarrow{\partial} \mathscr{D}^k(U) \xrightarrow{\partial} \mathscr{D}^{k-1}(U) \to \ldots$$

- $\mathrm{supp}(\partial T) \subseteq \mathrm{supp}(T)$ (since $\mathrm{supp}(\mathrm{d}\eta) \subseteq \mathrm{supp}(\eta)$);

- $T_i \rightharpoonup T \Longrightarrow \partial T_i \rightharpoonup \partial T$: indeed, given $\varphi \in \mathscr{D}^{k-1}(U)$ we have

$$\partial T_i(\varphi) = T_i(\mathrm{d}\varphi) \to T(\mathrm{d}\varphi) = \partial T(\varphi),$$

whenever $T_i \rightharpoonup T$. $\diamond$

The concept of a $k-$current on U is the measure-geometric generalization of the concept of an oriented $k-$submanifold on U with locally finite $k-$dimensional Hausdorff measure. This is the motivation for the definition of boundary for currents we have given above, as the following example illustrates:

Example A.59. For $k \geq 1$, let $N^k \subseteq U$ be an oriented $k-$submanifold with boundary ∂N in U and orientation ξ. Suppose $\mathcal{H}^k \lfloor N$ is locally finite (or, equivalently, a Radon measure). Then, N naturally induces a $k-$current $[[N]]$ on U given by

$$[[N]](\varphi) := \int_N \langle \varphi, \xi(x) \rangle \alpha_k \mathrm{d}\mathcal{H}^k = \int_N \varphi, \quad \forall \varphi \in \mathscr{D}^k(U).$$

Analogously, the boundary of N with the induced orientation induces a $(k-1)$–current $[[\partial N]]$ on U. Now, for each $\phi \in \mathscr{D}^{k-1}(U)$, by Stokes' theorem we have

$$[[\partial N]](\phi) = \int_{\partial N} \phi = \int_N \mathrm{d}\phi = [[N]](\mathrm{d}\phi) = \partial[[N]](\phi).$$

This shows that the boundary of the current determined by N, as per Definition A.57, equals the current determined by the boundary ∂N of N.

Definition A.60 (Total variation measure and mass of a current). Let $T \in \mathscr{D}_k(U)$. For an open subset $W \subseteq U$ and any $A \subseteq U$, we define

$$\|T\|(W) := \sup\{T(\varphi) : \mathrm{supp}(\varphi) \subseteq W, \ \|\varphi\|_{C^0} \leqslant 1\},$$

$$\|T\|(A) := \inf\{\|T\|(W) : A \subseteq W, \ W \text{ open}\}.$$

The resulting Borel regular (outer) measure $\|T\|$ is called the **total variational measure** of T. The (extended) number

$$\mathbf{M}(T) := \|T\|(U) = \sup\{T(\varphi) : \|\varphi\|_{C^0} \leqslant 1, \varphi \in \mathscr{D}^k(U)\} \in [0, \infty].$$

is called the **mass** of T.

Note that when $T = [[N]]$ is induced by a k–submanifold $N \subseteq U$ as in the above example, the total variation measure of T is simply $\|T\| = \mathcal{H}^k \lfloor N$, which shows that the mass generalizes the area of a submanifold.

Definition A.61 (Finite mass currents). Let $k \in \mathbb{N}_0$. We define the space $\mathbf{M}_k(U)$ of **finite mass** k–currents on U by

$$\mathbf{M}_k(U) := \{T \in \mathscr{D}_k(U) : \|T\| \text{ is a finite measure}\}.$$

$\mathbf{M}_k(U)$ has a natural structure of normed space induced by the **mass norm** $\mathbf{M}(T) := \|T\|$.

More generally, we define the space $\mathbf{M}_{k,\mathrm{loc}}(U)$ of **locally finite mass** k–currents on U by

$$\mathbf{M}_{k,\mathrm{loc}}(U) := \{T \in \mathscr{D}_k(U) : \|T\| \text{ is a Radon measure}\}.$$

$\mathbf{M}_{k,\mathrm{loc}}(U)$ has a natural topology induced by the family of semi-norms $\{\mathbf{M}_W\}_{W \Subset U}$, where $\mathbf{M}_W(T) := \|T\|(W)$.

A current $T \in \mathscr{D}_k(U)$ is said to be **representable by integration** when there exist a Radon measure μ_T over U and a μ_T–measurable function $\xi : U \to \Lambda^k \mathbb{R}^n$, with $|\xi| = 1$ μ_T–a.e., such that

$$T(\varphi) = \int_U \langle \varphi, \xi \rangle \mathrm{d}\mu_T.$$

In such case, one may prove that $\mu_T = \|T\|$. In particular, when T is representable by integration then $T \in \mathbf{M}_{k,\mathrm{loc}}(U)$. The converse follows from the Riesz representation theorem (Theorem A.24). Thus:

Theorem A.62 (Integral representation). *Let $T \in \mathscr{D}_k(U)$. Then, T is representable by integration if, and only if, $T \in \mathbf{M}_{k,\mathrm{loc}}(U)$.*

Moreover, we have the following standard weak* compactness result, which follows from the standard Banach–Alaoglu theorem (cf. Simon (1983, Lemma 26.14, p. 135)).

Lemma A.63. *Let $\{T_i\} \subseteq \mathbf{M}_{k,\mathrm{loc}}(U)$ be such that*

$$\sup_{i \geq 1} \|T_i\|(W) < \infty, \quad \text{for each } W \Subset U.$$

Then, after passing to a subsequence, there exists $T \in \mathbf{M}_{k,\mathrm{loc}}(U)$ such that $T_i \rightharpoonup T$.

Next we introduce various important spaces of currents.

Definition A.64 (Normal currents). Let $k \in \mathbb{N}_0$. We define the space $\mathbf{N}_k(U)$ of **normal** $k-$currents on U by

$$\mathbf{N}_k(U) := \{T \in \mathscr{D}_k(U) : \|T\| + \|\partial T\| \text{ is a finite measure}\}.$$

$\mathbf{N}_k(U)$ has a natural structure of normed space induced by $\mathbf{N}(T) := \mathbf{M}(T) + \mathbf{M}(\partial T)$.

More generally, we define the space $\mathbf{N}_{k,\mathrm{loc}}(U)$ of **locally normal** $k-$currents on U by

$$\mathbf{N}_{k,\mathrm{loc}}(U) := \{T \in \mathscr{D}_k(U) : \|T\| + \|\partial T\| \text{ is a Radon measure}\}.$$

$\mathbf{N}_{k,\mathrm{loc}}(U)$ has a natural topology induced by the family of semi-norms $\{\mathbf{N}_W\}_{W \Subset U}$, where $\mathbf{N}_W(T) := \mathbf{M}_W(T) + \mathbf{M}_W(\partial T)$.

Definition A.65 (Integer rectifiable currents). A $k-$current $T \in \mathscr{D}_k(U)$ is called **locally integer rectifiable** if there is a triple (Γ, ξ, Θ) such that:

(i) $\Gamma \subseteq U$ is $\mathcal{H}^k-$measurable and countably $\mathcal{H}^k-$rectifiable;

(ii) $\Theta \colon \Gamma \to [0, \infty[$ is locally $\mathcal{H}^k-$integrable and such that $\Theta(x) \in \mathbb{Z}$ for $\mathcal{H}^k-$a.e. $x \in \Gamma$;

(iii) $\xi \colon \Gamma \to \Lambda^k \mathbb{R}^n$ is $\mathcal{H}^k$–measurable and such that $\xi(x)$ *orients* the approximate k–tangent space $T_x\Gamma$ for $\mathcal{H}^k$–a.e. $x \in \Gamma$, that is, for $\mathcal{H}^k$–a.e. $x \in \Gamma$, $\xi(x) \in \Lambda^k \mathbb{R}^n$ is simple, unitary and represents the approximate k–tangent space $T_x\Gamma$;

(iv) the current T is given by

$$T(\varphi) := \int_\Gamma \langle \varphi, \xi \rangle \Theta \mathrm{d}\mathcal{H}^k, \quad \forall \varphi \in \mathscr{D}^k(U).$$

We call Θ the **multiplicity** of T and ξ the **orientation** of T; we write $T = (\Gamma, \xi, \Theta)$.

The set of *locally integer rectifiable k–currents* on U is denoted by $\mathbf{R}_{k,\mathrm{loc}}(U)$. The set of **integer rectifiable** k–currents on U is defined by

$$\mathbf{R}_k(U) := \mathbf{R}_{k,\mathrm{loc}}(U) \cap \mathbf{M}_k(U).$$

Remark A.66. In the literature, the space $\mathbf{R}_{k,\mathrm{loc}}(U)$ is sometimes simply called the space of *locally rectifiable k–currents* on U (note the missing of 'integer'). $\Diamond$

In general, if $T \in \mathbf{R}_{k,\mathrm{loc}}$ it need not be true that $\partial T \in \mathbf{R}_{k-1,\mathrm{loc}}$.

Definition A.67 (Integral currents). The space of **locally integral** k–currents on U is defined by

$$\mathbf{I}_{k,\mathrm{loc}}(U) := \{T \in \mathbf{R}_{k,\mathrm{loc}} : \partial T \in \mathbf{R}_{k-1,\mathrm{loc}}\}, \quad \text{if } k \geq 1,$$

and we set

$$\mathbf{I}_{0,\mathrm{loc}}(U) := \mathbf{R}_{0,\mathrm{loc}}.$$

The space $\mathbf{I}_k(U)$ of **integral** k–currents on U is defined by

$$\mathbf{I}_k(U) := \mathbf{I}_{k,\mathrm{loc}}(U) \cap \mathbf{N}_k(U).$$

The following theorem gives an important criterion for a k–current to be rectifiable Simon (1983, Theorem 32.1, pp. 183-187).

Theorem A.68 (Rectifiability Theorem). *If $T \in \mathscr{D}_k(U)$ is such that*

(i) $T \in \mathbf{N}_{k,\mathrm{loc}}(U)$ *(i.e. $\|T\|(W) + \|\partial T\|(W) < \infty$, $\forall W \Subset U$), and*

(ii) $\Theta^{*k}(\|T\|, x) > 0$ *for $\|T\|$–a.e. $x \in U$,*

then T is rectifiable, i.e. T is defined by a triple (Γ, ξ, Θ), in the sense that

$$T(\varphi) = \int_\Gamma \langle \varphi, \xi \rangle \Theta \, d\mathcal{H}^k \lfloor \Gamma, \quad \forall \varphi \in \mathscr{D}^k(U),$$

where

1. *$\Gamma \subseteq U$ is $\mathcal{H}^k$–measurable and countably $\mathcal{H}^k$–rectifiable;*

2. *$\Theta \colon \Gamma \to [0, \infty[$ is locally $\mathcal{H}^k$–integrable;*

3. *$\xi \colon \Gamma \to \Lambda^k \mathbb{R}^n$ is $\mathcal{H}^k$–measurable and such that $\xi(x)$ orients the approximate k–tangent space $T_x \Gamma$ for $\mathcal{H}^k$–a.e. $x \in \Gamma$.*

Definition A.69 (Cycles, boundaries, etc.). For $k \geqslant 1$, define the abelian groups

$$\mathcal{Z}_k(U) := \{T \in \mathbf{I}_k(U) : \partial T = 0\},$$
$$\mathcal{B}_k(U) := \{\partial S : S \in \mathbf{I}_{k+1}(U)\} \subseteq \mathcal{Z}_k(U).$$

An element of $\mathcal{Z}_k(U)$ is called a **cycle**; an element of $\mathcal{B}_k(U)$ is called a **boundary**. Two cycles $T, T' \in \mathbf{I}_k(U)$ are called **homologous** if $T - T'$ is a boundary.

Currents on manifolds. We now explain how the definition of currents on open subsets of Euclidean spaces can be transported to general manifolds. The key observation is the following. Let $F \colon U \to V$ be a coordinate change between two coordinate systems $(U, x^1, \ldots, x^n)$ and $(V, y^1, \ldots, y^n)$, where $U, V \subseteq \mathbb{R}^n$ are open subsets. Thus, F is a diffeomorphism such that $x^i = y^i \circ F$, for each $i = 1, \ldots, n$.

Claim. Let $F \colon U \to V$ be a diffeomorphism; then the natural induced map

$$F^* \colon \mathscr{D}^k(V) \to \mathscr{D}^k(U)$$

is an isomorphism of topological vector spaces.

Since F is a diffeomorphism, it is clear that F^* is a linear isomorphism. Moreover, since $(F^*)^{-1} = (F^{-1})^*$, in order to prove F^* is a homeomorphism it suffices to show that F^* is continuous (then the same argument will apply to the inverse map switching the roles of F and F^{-1}). By Proposition A.53, showing the continuity of F^*, in turn, boils down to proving the following: if $\phi^i \to 0$ in $\mathscr{D}^k(V)$ then $\varphi^i := F^*(\phi^i) \to 0$ in $\mathscr{D}^k(U)$.

Now suppose that $\phi^i \to 0$ in $\mathscr{D}^k(V)$. In particular, by Proposition A.52 (i), there exists a compact subset $\tilde{K} \subseteq V$ such that $\mathrm{supp}(\phi^i) \subseteq \tilde{K}$ for each i. It follows that the compact subset $K := F^{-1}(\widetilde{K}) \subseteq U$ is such that $\mathrm{supp}(\varphi^i) \subseteq K$, for each i. We write

$$\phi^i = \sum_J \phi^i_J \, \mathrm{d}y^J,$$

so that

$$\varphi^i = \sum_J \left(\phi^i_J \circ F\right) \mathrm{d}x^J =: \sum_J \varphi^i_J \mathrm{d}x^J.$$

Since

$$\frac{\partial}{\partial y^i} = F_* \frac{\partial}{\partial x^i},$$

it follows from Proposition A.52 (ii) that

$$\sup_{x \in K} \; |(D^\alpha \varphi^i_J)(x)| = \sup_{y \in \widetilde{K}} \; |(\widetilde{D}^\alpha \phi^i_J)(y)| \to 0 \quad \text{as } i \to \infty,$$

for all J and $\alpha \in \mathbb{N}_0^n$, where

$$D^\alpha := \left(\frac{\partial}{\partial x^1}\right)^{\alpha_1} \circ \ldots \circ \left(\frac{\partial}{\partial x^n}\right)^{\alpha_n}, \quad \text{and}$$

$$\widetilde{D}^\alpha := \left(\frac{\partial}{\partial y^1}\right)^{\alpha_1} \circ \ldots \circ \left(\frac{\partial}{\partial y^n}\right)^{\alpha_n}.$$

Therefore, by Proposition A.52, $\varphi^i := F^*(\phi^i) \to 0$ in $\mathscr{D}^k(U)$. This proves the continuity of F^*, completing the proof of the claim.

Now let M be a smooth $n-$manifold. Then, by making use of local charts on M, and the above observation, we get a well-defined $C^\infty-$topology on the space $\mathscr{D}^k(M)$ of **smooth compactly supported** $k-$**forms** on M. Thus we can define:

Definition A.70. A $k-$**current** T on M is an element of the topological dual $\mathscr{D}_k(M) := \left(\mathscr{D}^k(M)\right)'$, i.e. a continuous linear functional $T : \mathscr{D}^k(M) \to \mathbb{R}$.

$\mathscr{D}_k(M)$ endowed with the weak* topology is called the **space of** $k-$**currents on** M.

The previous definitions and results of this section are then obviously adapted to this context.

Partial differential operators and Sobolev spaces

B

In this brief appendix we collect some basic terminology and facts on partial differential operators and Sobolev spaces which are specially used in Sections 1.1, 1.4 and 3.1. The main references for this appendix are Wehrheim (2004) and Nicolaescu (2014-03-20).

B.1 Partial differential operators

In this section, we provide some definitions concerning (linear) partial differential operators (PDO) on manifolds. We follow the algebraic point of view of Nicolaescu's lecture notes Nicolaescu (ibid., Chapter 10).

Let $E_i \to M$ be a $\mathbb{K}-$vector bundle over a smooth manifold M, $i = 1, 2$. We start letting $\mathbf{Op}(E_1, E_2)$ be the natural $\mathbb{K}-$vector space whose underlying set is given by

$$\mathbf{Op}(E_1, E_2) := \{P : \Gamma(E_1) \to \Gamma(E_2) : P \text{ is } \mathbb{K} - \text{linear}\}.$$

In what follows, we will regard PDO's from sections of E_1 to sections of E_2 as elements of $\mathbf{Op}(E_1, E_2)$ that interact in a specific way with the $C^\infty(M, \mathbb{K})-$module structure of $\Gamma(E_1)$ and $\Gamma(E_2)$.

Let $\mathrm{Hom}(E_1, E_2)$ denote the space of vector bundle homomorphisms from E_1 to E_2, i.e. the space of all $P \in \mathbf{Op}(E_1, E_2)$ such that P is $C^\infty(M, \mathbb{K})$–linear. Then we can write

$$\mathrm{Hom}(E_1, E_2) = \{P \in \mathbf{Op}(E_1, E_2) : \mathrm{ad}(f)P = 0, \forall f \in C^\infty(M, \mathbb{K})\}$$
$$=: \ker \mathrm{ad},$$

where

$$\mathrm{ad}(f) \colon \mathbf{Op}(E_1, E_2) \to \mathbf{Op}(E_1, E_2)$$
$$P \mapsto [P, f] := P \circ f - f \circ P.$$

Here we are regarding f as the natural $C^\infty(M, \mathbb{K})$–module multiplication operator it induces on $\Gamma(E_1)$ and $\Gamma(E_2)$ where appropriate.

Definition B.1 (PDO's). Let $E_1, E_2 \to M$ be $\mathbb{K}$–vector bundles. For each $m \in \mathbb{N}_0$, we let $\mathbf{PDO}^{(m)}(E_1, E_2)$ be the set of all $P \in \mathbf{Op}(E_1, E_2)$ such that

$$\mathrm{ad}(f_0)\mathrm{ad}(f_1) \cdots \mathrm{ad}(f_m)P = 0, \ \forall f_i \in C^\infty(M, \mathbb{K})$$

and we set
$$\mathbf{PDO}(E_1, E_2) := \bigcup_{m \geq 0} \mathbf{PDO}^{(m)}(E_1, E_2).$$

An element $P \in \mathbf{PDO}(E_1, E_2)$ is called a **partial differential operator** from E_1 to E_2.

Definition B.2 (Formal adjoint). Suppose (M, g) is an oriented Riemannian manifold and let $E_i \to M$ be a $\mathbb{K}$–vector bundle over M endowed with a metric $\langle \cdot, \cdot \rangle_i$, $i = 1, 2$. Given $P \in \mathbf{PDO}(E_1, E_2)$, we say that $Q \in \mathbf{PDO}(E_2, E_1)$ is a **formal adjoint** of P whenever

$$\int_M \langle Pu, v \rangle_2 \mathrm{d}V_g = \int_M \langle u, Qv \rangle_1 \mathrm{d}V_g,$$

for each $u \in \Gamma(E_1)$ and $v \in \Gamma(E_2)$ one of which has compact support[1] in M.

Proposition B.3 (Existence and uniqueness of formal adjoints). *Suppose (M, g) is an oriented Riemannian manifold and let $E_i \to M$ be a $\mathbb{K}$–vector bundle over M endowed with a metric $\langle \cdot, \cdot \rangle_i$, $i = 1, 2$. Then for any $P \in$ $\mathbf{PDO}(E_1, E_2)$ there exists a unique formal adjoint $P^* \in \mathbf{PDO}(E_2, E_1)$ of P.*

---

[1] When $\partial M \neq \emptyset$, one assumes that the compact support lies in the *interior* of the manifold M.

B.2 Sobolev spaces

We now introduce Sobolev spaces of sections of *vector* bundles, and state the corresponding so-called Sobolev embedding theorems. This is the minimal background material to deal with Sobolev spaces of connections on G−bundles (see §1.1 of Chapter 1). For a definition of Sobolev spaces of sections of general fiber bundles[2], as well as other details, we refer the reader to Wehrheim (2004, Appendix B).

Let M be an oriented n−manifold endowed with a Riemannian metric g, and let $\pi\colon F \to M$ be a $\mathbb{K}$−vector bundle over M endowed with a metric $h = \langle\cdot,\cdot\rangle$ and associated pointwise norm $|\cdot|$. Henceforth, we use the notations introduced in Chapter 1.

Definition B.4 (L^p−sections). Let $1 \leqslant p \leqslant \infty$. We define the **Lebesgue space** $L^p(M, F)$ of L^p−sections of $F \to M$ to be the natural $\mathbb{K}$−vector space whose underlying set consists of all (equivalence classes, modulo the relation of equality μ_g−almost everywhere, of) Borel measurable maps $u\colon M \to F$ such that the following holds.

(i) $(\pi \circ u)(x) = x$, for μ_g−almost all $x \in M$.

(ii) The function $|u|\colon M \to \mathbb{R}$ defines an element in $L^p(\mu_g)$.

The L^p−**norm** $\|\cdot\|_p$ on $L^p(M, F)$ is given by

$$\|u\|_p := \begin{cases} \left(\int_M |u|^p \, dV_g \right)^{1/p}, & \text{if } 1 \leqslant p < \infty, \\ \operatorname*{ess\,sup}_M |s|, & \text{if } p = \infty. \end{cases}$$

More generally, we define the space $L^p_{\text{loc}}(M, F)$ of *locally* L^p−integrable sections of $F \to M$ by

$$L^p_{\text{loc}}(M, F) := \{u : fu \in L^p(M, F) \text{ for all } f \in C^\infty_c(M)\}.$$

Given an exhaustion $\{\Omega_i\}$ of M by precompact open subsets $\Omega_i \Subset M$, the space $L^p_{\text{loc}}(M, F)$ is endowed with the natural Fréchet topology induced by the family of semi-norms $\{\|\cdot\|_{p,\Omega_i}\}_{i\in\mathbb{N}}$ given by

$$\|u\|_{p,\Omega_i} := \int_{\Omega_i} |u|^p \, dV_g, \quad \forall u \in L^p_{\text{loc}}(M, F).$$

---

[2]This would cover, for instance, the case of Sobolev spaces of gauge transformations of G−bundles.

Remark B.5. Suppose $1 \leqslant p, q \leqslant \infty$ are *Hölder conjugate*, i.e. $1/p + 1/q = 1$, and let $u \in L^p(M, F)$ and $v \in L^q(M, F)$. Then, using the Cauchy–Schwarz inequality together with Hölder's inequality for functions, one gets

$$\int_M |\langle u, v \rangle| \mathrm{d}V_g \leqslant \|u\|_p \|v\|_q.$$

More generally, let $F_i \to M$ $(i = 1, \ldots, l)$ be vector bundles with metrics and consider $F_1^* \otimes \ldots \otimes F_l^*$ endowed with the induced tensor product metric. If $\Omega \in L^{p_0}(F_1^* \otimes \ldots \otimes F_l^*)$ for some $1 \leqslant p_0 \leqslant \infty$, then for every $1 \leqslant p_1, \ldots, p_l \leqslant \infty$ such that

$$1 - \frac{1}{p_0} = \frac{1}{p_1} + \ldots + \frac{1}{p_l},$$

and for every $u_i \in L^{p_i}(M, F_i)$, $i = 1, \ldots, l$, one can prove that

$$\int_M |\Omega(u_1, \ldots, u_l)| \, \mathrm{d}V_g \leqslant \|\Omega\|_{p_0} \|u_1\|_{p_1} \cdots \|u_l\|_{p_l}.$$

$\Diamond$

Lemma B.6. *The Lebesgue space $(L^p(M, F), \|\cdot\|_p)$ is a Banach space which is reflexive for $1 < p < \infty$.*

Now fix a smooth connection ∇ on F compatible with h. In what follows, we still denote by ∇ the tensor product connections (1.24) induced by ∇ and the Levi-Civita connection D^g of (M, g).

Definition B.7. Let $u \in L^1_{\mathrm{loc}}(M, F)$ and let $v \in L^1_{\mathrm{loc}}(M, \bigotimes^j T^*M \otimes F)$. We say that $\nabla^j u = v$ **weakly** if

$$\int_M \langle u, (\nabla^j)^* \phi \rangle \mathrm{d}V_g = \int_M \langle v, \phi \rangle \mathrm{d}V_g, \quad \forall \phi \in \Gamma_0(\overset{j}{\bigotimes} T^*M \otimes F),$$

where $(\nabla^j)^*$ denotes the formal adjoint of $\nabla^j \in \mathbf{PDO}^{(j)}(F, \bigotimes^j T^*M \otimes F)$.

Definition B.8 ($W^{k,p}$–sections). Let $1 \leqslant p \leqslant \infty$ and let $k \in \mathbb{N}_0$. We define the **Sobolev space** $W^{k,p}(M, F)$ of $W^{k,p}$–sections of $F \to M$ to be the natural $\mathbb{K}$–vector space whose underlying set consists of all $u \in L^p(M, F)$ such that, for each $1 \leqslant j \leqslant k$, there exists $v_j \in L^p(M, \bigotimes^j T^*M \otimes F)$ satisfying $\nabla^j u = v_j$ weakly. The **Sobolev $W^{k,p}$–norm** $\|\cdot\|_{p,k}$ on $W^{k,p}(M, F)$ is given by

$$\|u\|_{k,p} := \sum_{j=0}^k \|\nabla^j u\|_p.$$

Note that, by definition, $W^{0,p}(M, F) = L^p(M, F)$. Moreover:

Theorem B.9. *$W^{k,p}(M, F)$ is a Banach space which is reflexive for $1 < p < \infty$.*

By the Banach–Alaoglu theorem of functional analysis, one gets:

Corollary B.10. *If $1 < p < \infty$, then every bounded sequence in $W^{k,p}(M, F)$ has a weakly convergent subsequence.*

Let $\Gamma_0(F)$ denote the space of *compactly supported* sections of $F \to M$. The next result implies that we could have defined $W^{k,p}(M, F)$ as the norm completion of $(\Gamma_0(F), \| \cdot \|_{k,p})$.

Proposition B.11. *If $1 \leqslant p < \infty$, then $\Gamma_0(F)$ is dense on $W^{k,p}(M, F)$.*

Contrary to what our notation suggests so far, the spaces $W^{k,p}(M, F)$ may heavily depend on the choices of a metric g on M, a metric h on E and, in case $k \geqslant 1$, the choice of a compatible connection ∇ on F. In fact, when M is noncompact, this dependence has to be seriously taken into account. On the other hand, it turns out that for *compact* base manifolds M these spaces are independent of these choices and, although their norms always depend on the various choices of g, h and (possibly) ∇ (all of which will be clear in the context), a change of choices always gives equivalent norms. Indeed, we have the following (cf. Nicolaescu (2014-03-20, p.251, Theorem 10.2.36)):

Theorem B.12. *Let $F \to M$ be a vector bundle over a compact, oriented, $n-$manifold M. Suppose that g_i is a Riemannian metric on M, h_i is a metric on F and that ∇_i is a smooth connection on F compatible with h_i, where $i = 1, 2$. Then we have the set equality*

$$W^{k,p}(M, F; g_1, h_1, \nabla_1) = W^{k,p}(M, F; g_2, h_2, \nabla_2)$$

and the identity map between these Banach spaces is a bounded linear map.

We finish this appendix stating the so-called Sobolev embeddings (cf. Wehrheim (2004, Theorem B.2, p. 182)). In what follows, we suppose M to be a *compact*, oriented, $n-$manifold. Moreover, for each $j \in \mathbb{N}_0$, we let $C^j(M, F)$ be the space of C^j-sections of $F \to M$, i.e. the space of all maps $u \colon M \to F$ of class C^j such that $\pi \circ u = \mathbb{1}_M$. We endow $C^j(M, F)$ with the uniform C^j-topology induced by the $W^{j,\infty}-$norm.

Theorem B.13 (Sobolev embeddings)**.** *Let $0 \leqslant j < k$ be integers and let $1 \leqslant p, q < \infty$ be real numbers.*

(i) *If $k - \frac{n}{p} \geqslant j - \frac{n}{q}$ then the natural inclusion*

$$W^{k,p}(M, F) \hookrightarrow W^{j,q}(M, F)$$

is a bounded linear map. Moreover, if strictly inequality holds this inclusion map is compact.

(ii) *If $k - \frac{n}{p} > j$ then there is a compact bounded inclusion map*

$$W^{k,p}(M, F) \hookrightarrow C^j(M, F).$$

As a consequence, we have the following multiplication theorem.

Theorem B.14. *Let $k \in \mathbb{N}_0$ and let $1 \leqslant p, r, s < \infty$ be such that either*

$$r, s \geqslant p \quad and \quad \frac{1}{r} + \frac{1}{s} < \frac{k}{n} + \frac{1}{p}$$

or

$$r, s > p \quad and \quad \frac{1}{r} + \frac{1}{s} \leqslant \frac{k}{n} + \frac{1}{p}.$$

Then, the (pointwise) multiplication map

$$W^{k,r}(M) \times W^{k,s}(M) \to W^{k,p}(M)$$

is well-defined and continuous.

D. V. Alekseevsky, V. Cortés, and C. Devchand (2003). "Yang–Mills connections over manifolds with Grassmann structure." *Journal of Mathematical Physics* 44.12, pp. 6047–6076. MR: 2023568 (cit. on p. 77).

F. J. Almgren (1984). "$Q-$valued functions minimizing Dirichlet's integral and the regularity of area minimizing rectifiable currents up to codimension two." *Princeton, NJ* (cit. on p. 68).

M. F. Atiyah, N. J. Hitchin, V. G. Drinfeld, and Y. I. Manin (1978). "Construction of instantons." *Physics Letters A* 65, pp. 185–187. MR: 0598562 (cit. on p. 43).

T. Aubin (1982). *Nonlinear Analysis on Manifolds. Monge–Ampère Equations*. Vol. 252. Springer Science & Business Media. MR: 0681859 (cit. on pp. 101, 145).

G. Ball and G. Oliveira (2019). "Gauge theory on Aloff–Wallach spaces." *Geometry and Topology* 23. MR: 3939051 (cit. on p. 86).

L. Baulieu, H. Kanno, and I. M. Singer (1998). "Special quantum field theories in eight and other dimensions." *Communications in Mathematical Physics* 194.1, pp. 149–175. MR: 1628298 (cit. on pp. 2, 6, 49, 73).

A. A. Belavin, A. M. Polyakov, A. S. Schwartz, and Y. S. Tyupkin (1975). "Pseudoparticle solutions of the Yang–Mills equations." *Physics Letters B* 59.1, pp. 85–87. MR: 0434183 (cit. on p. 41).

M. Berger (1955). "Sur les groupes d'holonomie homogènes de variétés à connexion affine et des variétés riemanniennes." *Bulletin de la Société Mathématique de France* 83, pp. 279–330. MR: 0079806 (cit. on p. 51).

A. L. Besse (2008). *Einstein manifolds. Reprint of the 1987 edition. Classics in Mathematics*. MR: 2371700 (cit. on p. 37).

J. P. Bourguignon and H. B. Lawson Jr (1981). "Stability and isolation phenomena for Yang–Mills fields." *Communications in Mathematical Physics* 79.2, pp. 189–230. MR: 0612248 (cit. on pp. 6, 22).

R. L. Bryant (1986). "A survey of Riemannian metrics with special holonomy groups." In: *Proceedings of the International Congress of Mathematicians*. Vol. 1, pp. 505–514. MR: 0934250 (cit. on pp. 50, 51, 55).

— (1987). "Metrics with exceptional holonomy." *Annals of Mathematics*, pp. 525–576. MR: 0916718 (cit. on pp. 55, 56, 58, 59).

R. L. Bryant and S. Salamon (1989). "On the construction of some complete metrics with exceptional holonomy." *Duke Mathematical Journal* 58.3, pp. 829–850. MR: 1016448 (cit. on pp. 58, 85, 88).

D. Burago, Y. Burago, and S. Ivanov (2001). *A course in metric geometry*. Vol. 33. American Mathematical Society Providence. MR: 1835418 (cit. on p. 146).

R. R. Carrión (1998). "A generalization of the notion of instanton." *Differential Geometry and its Applications* 8.1, pp. 1–20. MR: 1601546 (cit. on pp. 2, 6, 49, 79).

A. Clarke (2014). "Instantons on the exceptional holonomy manifolds of Bryant and Salamon." *Journal of Geometry and Physics* 82, pp. 84–97. MR: 3206642 (cit. on pp. 85, 88).

A. Clarke and G. Oliveira (2019). "Spin(7)-Instantons from evolution equations." arXiv: 1903.05526 (cit. on pp. 8, 86, 136).

A. Clarke and B. Santoro (2012). "Holonomy groups in Riemannian geometry." arXiv: 1206.3170 (cit. on p. 24).

E. Corrigan, C. Devchand, D. B. Fairlie, and J. Nuyts (1983). "First–order equations for gauge fields in spaces of dimension greater than four." *Nuclear Physics B* 214.3, pp. 452–464. MR: 0698892 (cit. on pp. 2, 6, 49, 73).

A. Corti, M. Haskins, J. Nordström, and T. Pacini (2013). "Asymptotically cylindrical Calabi–Yau 3–folds from weak Fano 3–folds." *Geometry & Topology* 17.4, pp. 1955–2059. MR: 3109862 (cit. on p. 54).

— (2015). "G_2–manifolds and associative submanifolds via semi–Fano 3–folds." *Duke Mathematical Journal* 164.10, pp. 1971–2092. MR: 3369307 (cit. on pp. 59, 71).

C. De Lellis (2008). *Rectifiable sets, densities and tangent measures*. European Mathematical Society. MR: 2388959 (cit. on pp. 139, 152, 156).

A. Doan (2019). "Seiberg–Witten monopoles with multiple spinors on a surface times a circle." *Journal of Topology* 12.1, pp. 1–55. MR: 3875977 (cit. on p. 137).

A. Doan and T. Walpuski (2017). "On counting associative submanifolds and Seiberg–Witten monopoles." arXiv: 1712.08383 (cit. on p. 137).

S. K. Donaldson (1983). "An application of gauge theory to four dimensional topology." *Journal of Differential Geometry* 18.2, pp. 279–315. MR: 0710056 (cit. on p. 2).

— (1985). "Anti self–dual Yang–Mills connections over complex algebraic surfaces and stable vector bundles." *Proceedings of the London Mathematical Society* 50.1, pp. 1–26. MR: 0765366 (cit. on p. 81).

— (1990). "Polynomial invariants for smooth four–manifolds." *Topology* 29.3, pp. 257–315. MR: 1066174 (cit. on p. 2).

— (2005). "Yang–Mills Theory and Geometry" (cit. on p. 2).

S. K. Donaldson and P. B. Kronheimer (1990). *The geometry of four–manifolds*. Clarendon Press. MR: 1079726 (cit. on pp. 2, 10, 28, 37, 43, 45–47, 96, 97, 130).

S. K. Donaldson and E. P. Segal (2011). "Gauge theory in higher dimensions, II." In: *Surveys in differential geometry. Volume XVI. Geometry of special holonomy and related topics*. Vol. 16. Surv. Differ. Geom. Int. Press, Somerville, MA, pp. 1–41. MR: 2893675 (cit. on pp. 49, 136).

S. K. Donaldson and R. P. Thomas (1998). "Gauge theory in higher dimensions." *The Geometric Universe (Oxford, 1996)*, pp. 31–47. MR: 1634503 (cit. on pp. 1, 2, 6, 49, 88, 136).

J. Eichhorn (1991). "The boundedness of connection coefficients and their derivatives." *Mathematische Nachrichten* 152.1, pp. 145–158. MR: 1121230 (cit. on p. 102).

D. G. Fadel (2016). "On blow-up loci of instantons in higher dimensions." MA thesis. SP, Brazil: University of Campinas (Unicamp) (cit. on p. 1).

H. Federer (1969). *Geometric measure theory*. Springer. MR: 0257325 (cit. on pp. 139, 144).

— (1978). "Colloquium lectures on geometric measure theory." *Bulletin of the American Mathematical Society* 84.3, pp. 291–338. MR: 0467473 (cit. on p. 160).

M. Fernández and A. Gray (1982). "Riemannian manifolds with structure group G_2." *Annali di Matematica Pura ed Applicata* 132.1, pp. 19–45. MR: 0696037 (cit. on p. 57).

G. B. Folland (2013). *Real analysis: modern techniques and their applications*. John Wiley & Sons. MR: 1681462 (cit. on pp. 34, 118, 139, 141, 152).

D. S. Freed and K. K. Uhlenbeck (1984). *Instantons and four–manifolds*. Springer. MR: 0757358 (cit. on pp. 2, 10, 37).

M. H. Freedman (1982). "The topology of four–dimensional manifolds." *Journal of Differential Geometry* 17.3, pp. 357–453. MR: 0679066 (cit. on p. 2).

D. Gilbarg and N. S. Trudinger (2001). *Elliptic partial differential equations of second order*. Vol. 224. Springer. MR: 1814364 (cit. on p. 109).

R. E. Gompf (1985). "An infinite set of exotic $\mathbb{R}^4$'s." *Journal of Differential Geometry* 21.2, pp. 283–300. MR: 0816673 (cit. on p. 2).

D. Gromoll and G. Walschap (2009). *Metric foliations and curvature*. Vol. 268. Springer Science & Business Media. MR: 2500106 (cit. on p. 102).

R. Harvey and H. B. Lawson (1982). "Calibrated geometries." *Acta Mathematica* 148.1, pp. 47–157. MR: 0666108 (cit. on pp. 1, 6, 61, 64, 70, 72).

R. Harvey and B. Shiffman (1974). "A characterization of holomorphic chains." *Annals of Mathematics*, pp. 553–587. MR: 0355095 (cit. on p. 135).

M. Haskins, H.-J. Hein, and J. Nordström (2015). "Asymptotically cylindrical Calabi–Yau manifolds." *Journal of DIfferential Geometry* 101.2, pp. 213–265. MR: 3399097 (cit. on p. 59).

A. Haydys (2012). "Gauge theory, calibrated geometry and harmonic spinors." *Journal of the London Mathematical Society* 86.2, pp. 482–498. MR: 2980921 (cit. on p. 137).

— (2017). "G2 instantons and the Seiberg–Witten monopoles." arXiv: 1703.06329 (cit. on pp. 6, 137).

A. Haydys and T. Walpuski (2015). "A compactness theorem for the Seiberg–Witten equation with multiple spinors in dimension three." *Geometric and Functional Analysis* 25.6, pp. 1799–1821. MR: 3432158 (cit. on p. 137).

E. Hebey (2000). *Nonlinear analysis on manifolds: Sobolev spaces and inequalities*. Vol. 5. American Mathematical Society. MR: 1688256 (cit. on p. 103).

S. Hohloch, G. Noetzel, and D. A. Salamon (2009). "Hypercontact structures and Floer homology." *Geometry & Topology* 13.5, pp. 2543–2617. MR: 2529942 (cit. on p. 109).

A. Jacob and T. Walpuski (2018). "Hermitian Yang–Mills metrics on reflexive sheaves over asymptotically cylindrical Kähler manifolds." *Communications in Partial Differential Equations* 43.11, pp. 1566–1598. MR: 3924216 (cit. on pp. 8, 85).

M. Jardim (2005). *An introduction to gauge theory and its applications*. IMPA. MR: 2166639 (cit. on p. 40).

M. Jardim, G. Menet, D. M. Prata, and H. N. Sá Earp (2017). "Holomorphic bundles for higher dimensional gauge theory." *Bulletin of the London Mathematical Society* 49.1. MR: 3653106 (cit. on p. 85).

D. D. Joyce (1996). "Compact Riemannian 7–manifolds with holonomy G_2. I, II." *Journal of Differential Geometry* 43, pp. 291–328, 329–375. MR: 1424428 (cit. on pp. 59, 85, 88).

— (1999). "A new construction of compact 8-manifolds with holonomy Spin(7)." *Journal of Differential Geometry* 53.1, pp. 89–130. MR: 1776092 (cit. on p. 88).

— (2000). *Compact manifolds with special holonomy*. Oxford University Press. MR: 1787733 (cit. on pp. 50, 53, 59, 85).

— (2006). "The exceptional holonomy groups and calibrated geometry." In: *Proceedings of the Gökova Geometry-Topology Conference 2005* (Gökova, Turkey). Ed. by S. Akbulut, T. Onder, and R. Stern. International Press, pp. 110–139. MR: 2282012 (cit. on pp. 24, 50).

— (2007). *Riemannian Holonomy Groups and Calibrated Geometry*. Oxford University Press. MR: 2292510 (cit. on pp. 24, 26, 27, 50–52, 54, 55, 58, 60, 61, 70, 71).

— (2017). "Conjectures on counting associative 3-folds in G_2-manifolds." arXiv: 1610.09836 (cit. on p. 137).

D. D. Joyce and S. Karigiannis (2017). "A new construction of compact torsion-free G_2-manifolds by gluing families of Eguchi-Hanson spaces." to appear in Journal of Differential Geometry. arXiv: 1707.09325 (cit. on p. 59).

S. Kobayashi (2014). *Differential geometry of complex vector bundles*. Princeton University Press. MR: 3643615 (cit. on p. 81).

A. A. Kosinski (2007). *Differential manifolds*. Courier Corporation. MR: 1190010 (cit. on p. 26).

A. Kovalev (2003). "Twisted connected sums and special Riemannian holonomy." arXiv: math/0012189 (cit. on p. 59).

A. Kovalev and N.-H. Lee (2011). "K3 surfaces with non-symplectic involution and compact irreducible G_2-manifolds." In: *Mathematical Proceedings of the Cambridge Philosophical Society*. Vol. 151. 2. Cambridge University Press, pp. 193–218. MR: 2823130 (cit. on p. 59).

U. Lang (2007). "Notes on rectifiability" (cit. on p. 156).

H. B. Lawson (1980). "Lectures on minimal submanifolds." 1. MR: 0576752 (cit. on pp. 61, 63, 69).

C. Lewis (1999). "Spin(7)$-$instantons." PhD thesis. University of Oxford (cit. on p. 88).

F. H. Lin (1999). "Gradient estimates and blow–up analysis for stationary harmonic maps." *Annals of Mathematics* 149, pp. 785–829. MR: 1709303 (cit. on pp. 3, 7, 111, 112).

J. D. Lotay (2014). "Calibrated submanifolds" (cit. on p. 61).

J. D. Lotay and G. Oliveira (2018). "SU(2)^2-invariant G_2-instantons." *Mathematische Annalen* 371.1-2, pp. 961–1011. MR: 3788869 (cit. on pp. 8, 86, 136).

M. Lübke and A. Teleman (1995). *The Kobayashi–Hitchin correspondence*. World Scientific. MR: 1370660 (cit. on p. 81).

I. H. Madsen and J. Tornehave (1997). *From calculus to cohomology: de Rham cohomology and characteristic classes*. Cambridge University Press. MR: 1454127 (cit. on p. 16).

M. G. Martino (2011). "Gauge theory in dimensions two and four." MA thesis. Universidade Estadual de Campinas (cit. on p. 40).

P. Mattila and K. Falconer (1996). "Geometry of sets and measures in Euclidean spaces." *Bulletin of the London Mathematical Society* 28.134, pp. 544–545. MR: 0949087 (cit. on p. 139).

G. Menet, J. Nordström, and H. N. Sá Earp (2015). "Construction of G_2-instantons via twisted connected sums." arXiv: 1510.03836 (cit. on p. 85).

Y. Meyer and T. Rivière (2003). "A partial regularity result for a class of stationary Yang–Mills fields in high dimension." *Revista matemática iberoamericana* 19.1, pp. 195–219. MR: 1993420 (cit. on p. 136).

J. W. Milnor (1974). *Characteristic classes*. 76. Princeton University Press. MR: 0440554 (cit. on pp. 28, 29, 31).

A. Naber and D. Valtorta (2016). "Energy identity for stationary Yang Mills." arXiv: 1610.02898 (cit. on pp. 8, 130).

G. L. Naber (2011). *Topology, geometry, and gauge fields: Foundations*. Vol. 25. 2. Springer-Verlag New York. MR: 2676214 (cit. on p. 40).

H. Nakajima (1988). "Compactness of the moduli space of Yang–Mills connections in higher dimensions." *Journal of the Mathematical Society of Japan* 40.3, pp. 383–392. MR: 0945342 (cit. on pp. 3, 7, 91, 107, 108, 111).

L. I. Nicolaescu (2014-03-20). "Lectures on the geometry of manifolds" (cit. on pp. 167, 171).

J. Nordström (2012). "Calibrated Geometry" (cit. on p. 61).

C. Olmos (2005). "A geometric proof of the Berger holonomy theorem." *Annals of Mathematics*, pp. 579–588. MR: 2150392 (cit. on p. 52).

T. Pacini (2013). "Desingularizing isolated conical singularities: Uniform estimates via weighted Sobolev spaces." *Communications in Analysis and Geometry* 21.1, pp. 105–170. MR: 3046940 (cit. on p. 102).

P. Petersen (2006). *Riemannian geometry*. Vol. 171. Springer Science & Business Media. MR: 2243772 (cit. on p. 37).

M. Petrache and T. Rivière (2017). "The resolution of the Yang–Mills Plateau problem in super-critical dimensions." *Advances in Mathematics* 316, pp. 469–540. MR: 3672911 (cit. on pp. 8, 135).

— (2018). "The Space of Weak Connections in High Dimensions." arXiv: 1812.04432 (cit. on p. 136).

D. Preiss (1987). "Geometry of measures in $\mathbb{R}^n$: distribution, rectifiability, and densities." *Annals of Mathematics*, pp. 537–643. MR: 0890162 (cit. on p. 158).

P. Price (1983). "A monotonicity formula for Yang–Mills fields." *Manuscripta Mathematica* 43.2, pp. 131–166. MR: 0707042 (cit. on pp. 3, 7, 91, 97, 101, 106).

T. Rivière (2002). "Interpolation spaces and energy quantization for Yang–Mills fields." *Communications in Analysis and Geometry* 10.4, pp. 683–708. MR: 1925499 (cit. on p. 130).

W. Rudin (1986). *Real and complex analysis (3rd)*. New York: McGraw-Hill Inc. MR: 0344043 (cit. on p. 152).

— (1991). *Functional analysis. International series in pure and applied mathematics*. MR: 1157815 (cit. on pp. 159, 160).

H. N. Sá Earp (2009). "Instantons on G_2−manifolds." PhD thesis. Imperial College London (cit. on p. 85).

— (2015). "G_2−instantons over asymptotically cylindrical manifolds." *Geometry & Topology* 19.1, pp. 61–111. MR: 3318748 (cit. on pp. 8, 84, 85).

— (2018). "Current progress on G_2–instantons over twisted connected sums." to appear in a forthcoming volume of the Fields Institute Communications, entitled "Lectures and Surveys on G2 manifolds and related topics". arXiv: 1812.04664 (cit. on p. 85).

H. N. Sá Earp and T. Walpuski (2015). "G_2−instantons over twisted connected sums." *Geometry & Topology* 19.3, pp. 1263–1285. MR: 3352236 (cit. on pp. 8, 84, 85, 136).

L. Sadun and J. Segert (1991). "Non–self–dual Yang–Mills connections with nonzero Chern number." *American Mathematical Society* 24.1. MR: 1067574 (cit. on p. 39).

D. A. Salamon and T. Walpuski (2017). "Notes on the octonions." In: *Proceedings of the 23rd Gökova Geometry–Topology Conference*, pp. 1–85. MR: 3676083 (cit. on pp. 55, 82, 86).

S. Salamon (1989). *Riemannian geometry and holonomy groups*. Longman Scientific and Technical. MR: 1004008 (cit. on pp. 52, 57, 60, 77, 78, 82).

R. M. Schoen (1984). "Analytic aspects of the harmonic map problem." In: *Seminar on nonlinear partial differential equations*. Springer, pp. 321–358. MR: 0765241 (cit. on p. 107).

A. Scorpan (2005). *The wild world of 4−manifolds*. American Mathematical Society. MR: 2136212 (cit. on pp. 2, 37, 46).

L. M. Sibner, R. J. Sibner, and K. K. Uhlenbeck (1989). "Solutions to Yang–Mills equations that are not self–dual." *Proceedings of the National Academy of Sciences* 86.22, pp. 8610–8613. MR: 1023811 (cit. on p. 39).

L. Simon (1983). "Lectures on geometric measure theory, volume 3 of Proceedings of the Centre for Mathematical Analysis, Australian National University." *Australian National University Centre for Mathematical Analysis, Canberra* 6. MR: 0756417 (cit. on pp. 139, 142, 144, 146, 148, 150, 151, 154–156, 163, 164).

J. Simons (1962). "On the transitivity of holonomy systems." *Annals of Mathematics*, pp. 213–234. MR: 0148010 (cit. on p. 52).

Y. Tanaka (2012). "A construction of Spin(7)−instantons." *Annals of Global Analysis and Geometry* 42.4, pp. 495–521. MR: 2995202 (cit. on p. 88).

T. Tao and G. Tian (2004). "A singularity removal theorem for Yang–Mills fields in higher dimensions." *Journal of the American Mathematical Society* 17.3, pp. 557–593. MR: 2053951 (cit. on pp. 5, 136).

C. H. Taubes (1982). "Self–dual Yang–Mills connections on non self dual 4–manifolds." *Journal of Differential Geometry* 17.1, pp. 139–170. MR: 0658473 (cit. on p. 2).

— (1987). "Gauge theory on asymptotically periodic 4–manifolds." *Journal of Differential Geometry* 25.3, pp. 363–430. MR: 0882829 (cit. on p. 2).

R. P. Thomas (1997). "Gauge theory on Calabi–Yau manifolds." PhD thesis. University of Oxford (cit. on pp. 88, 136).

G. Tian (2000). "Gauge theory and calibrated geometry, I." *Annals of Mathematics* 151.1, pp. 193–268. MR: 1745014 (cit. on pp. 1, 3, 6–8, 49, 67, 73, 74, 91, 97, 99, 103, 105, 107, 111, 112, 115, 116, 118, 122, 124, 126, 129, 131, 132).

K. K. Uhlenbeck (1982a). "Connections with L^p bounds on curvature." *Communications in Mathematical Physics* 83.1, pp. 31–42. MR: 0648356 (cit. on pp. 2, 7, 24, 36, 91–93).

— (1982b). "Removable singularities in Yang–Mills fields." *Communications in Mathematical Physics* 83.1, pp. 11–29. MR: 0648355 (cit. on pp. 2, 7, 32, 35, 36, 91, 107, 123).

— (n.d.). "A priori estimates for Yang–Mills fields" (cit. on pp. 3, 107).

K. K. Uhlenbeck and S. T. Yau (1986). "On the existence of Hermitian–Yang–Mills connections in stable vector bundles." *Communications on Pure and Applied Mathematics* 39.S1, S257–S293. MR: 0861491 (cit. on pp. 81, 107).

T. Walpuski (2013a). "G_2–instantons on generalised Kummer constructions." *Geometry & Topology* 17.4, pp. 2345–2388. MR: 3110581 (cit. on pp. 85, 136).

— (2013b). "Gauge theory on G_2–manifolds." PhD thesis. Imperial College London (cit. on pp. 55, 73, 82, 85, 86, 137).

— (2016). "G_2-instantons over twisted connected sums: an example." *Mathematical Research Letters* 23.2, pp. 529–544. MR: 3512897 (cit. on p. 85).

— (2017a). "G_2–instantons, associative submanifolds and Fueter sections." *Communications in Analysis and Geometry* 25.4, pp. 847–893. MR: 3731643 (cit. on pp. 6, 136).

— (2017b). "Spin(7)–instantons, Cayley submanifolds and Fueter sections." *Communications in Mathematical Physics* 352.1, pp. 1–36. MR: 3623252 (cit. on pp. 6, 88, 136).

— (2017c). "A compactness theorem for Fueter sections." *Commentarii Mathematici Helvetici* 92.4, pp. 751–776. MR: 3718486 (cit. on pp. 7, 8, 109, 112, 115, 124).

H. Y. Wang (1991). "The existence of nonminimal solutions to the Yang–Mills equation with group SU(2) on $S^2 \times S^2$ and $S^1 \times S^3$." *Journal of Differential Geometry* 34.3, pp. 701–767. MR: 1139645 (cit. on p. 39).

F. W. Warner (2013). *Foundations of differentiable manifolds and Lie groups.* Vol. 94. Springer Science & Business Media. MR: 0722297 (cit. on p. 143).

K. Wehrheim (2004). *Uhlenbeck compactness.* European Mathematical Society. MR: 2030823 (cit. on pp. 23, 32, 35, 92–94, 96, 167, 169, 171).

S. T. Yau (1978). "On the Ricci curvature of a compact Kähler manifold and the complex Monge–Ampére equation, I." *Communications on Pure and Applied Mathematics* 31.3, pp. 339–411. MR: 0480350 (cit. on p. 55).

Títulos Publicados — 32º Colóquio Brasileiro de Matemática

www.ingramcontent.com/pod-product-compliance
Lightning Source LLC
LaVergne TN
LVHW020327200726
843507LV00012B/2275